SpringerBriefs in Electrical and Computer Engineering

Series Editors

Woon-Seng Gan, School of Electrical and Electronic Engineering, Nanyang Technological University, Singapore, Singapore

C.-C. Jay Kuo, University of Southern California, Los Angeles, USA

Thomas Fang Zheng, Research Institute of Information Technology, Tsinghua University, Beijing, China

Mauro Barni, Department of Information Engineering and Mathematics, University of Siena, Siena, Italy

SpringerBriefs present concise summaries of cutting-edge research and practical applications across a wide spectrum of fields. Featuring compact volumes of 50 to 125 pages, the series covers a range of content from professional to academic. Typical topics might include: timely report of state-of-the art analytical techniques, a bridge between new research results, as published in journal articles, and a contextual literature review, a snapshot of a hot or emerging topic, an in-depth case study or clinical example and a presentation of core concepts that students must understand in order to make independent contributions.

Martin Ćalasan · Snežana Vujošević

Analytical Solutions of Nonlinear Power System Models Using the Lambert W Function

A Research Monograph

 Springer

Martin Ćalasan
Faculty of Electrical Engineering
University of Montenegro
Podgorica, Montenegro

Snežana Vujošević
Faculty of Electrical Engineering
University of Montenegro
Podgorica, Montenegro

ISSN 2191-8112   ISSN 2191-8120 (electronic)
SpringerBriefs in Electrical and Computer Engineering
ISBN 978-3-032-09578-7   ISBN 978-3-032-09579-4 (eBook)
https://doi.org/10.1007/978-3-032-09579-4

This Springer imprint is published by the registered company Springer Nature Switzerland AG
The registered company address is: Gewerbestrasse 11, 6330 Cham, Switzerland

If disposing of this product, please recycle the paper.

Preface

The study of nonlinear models in electrical engineering has always represented a significant analytical and numerical challenge. In many practical applications, the behavior of devices and systems cannot be adequately described using simple linear formulations. Instead, engineers and researchers must rely on nonlinear formulations, which often lack closed-form analytical solutions. Among the mathematical tools that have recently gained importance in this field, the Lambert W function has emerged as a powerful means of obtaining exact or semi-exact solutions to a wide range of nonlinear problems.

This research monograph, *Analytical Solutions of Nonlinear Power System Models Using the Lambert W Function*, is the result of years of research in the field of modeling, analysis, and simulation of various power system components. Although the Lambert W function was introduced long ago, it has found widespread application in engineering and applied sciences only in the past three decades. Its unique ability to provide closed-form solutions to transcendental equations makes it particularly suitable for problems in electrical engineering involving diodes, solar cells, fuel cells, rectifiers, induction machines, and inductive elements.

The monograph is structured to first introduce the Lambert W function, its fundamental properties, and numerical methods for its computation. It then presents chapters focused on concrete applications of this function, covering areas such as modeling of solar cells, perovskite solar cells, dynamics of induction machines, proton-exchange membrane fuel cells, diode–resistor circuits, rectifier systems, and modeling of the air gap in inductors. Each application is accompanied by detailed mathematical derivations, analytical solutions, and MATLAB codes, enabling both theoretical understanding and practical implementation.

The purpose of this monograph is twofold: to serve as a comprehensive reference for researchers in the field of nonlinear modeling in electrical engineering, and to provide a practical guide for engineers seeking to apply analytical methods in solving complex problems in power systems. By combining mathematical rigor with engineering relevance, this monograph seeks to bridge the gap between theory and practice.

We hope that this book will encourage further exploration of the Lambert W function and related special functions in electrical engineering, inspire new research directions, and provide valuable tools for both the academic and industrial communities.

Podgorica, Montenegro

Martin Ćalasan
Snežana Vujošević

Acknowledgments

I would like to express my heartfelt gratitude to my wife, Sanja, whose unwavering support has been the cornerstone of my research journey. Her patience, understanding, and steadfast encouragement have guided me through every challenge, while her belief in my work has inspired me to pursue new ideas with determination. Truly, as the saying goes: "Behind every successful man stands a woman who believes in him."

—Prof. Dr. Martin Ćalasan, Associate Professor

I am especially grateful to my children for their support and understanding during my work of this book.

—Dr. Snežana Vujošević, Assistant Professor

Contents

Chapter 1
Lambert W Function

This chapter introduces the Lambert W function and its extended form, the g-function (LogWright function). Historical development, mathematical properties, and practical applications of these functions in engineering and physics are discussed. Both numerical and analytical methods for solving Lambert W and g-function equations are presented, including Newton's and Halley's iterative procedures, Taylor series expansions, and the Special Trans Function Theory (STFT). Numerical examples illustrate the accuracy and convergence of these methods.

1.1 Historical Background and Definition

Johann Heinrich Lambert was a German-Swiss scientist born in 1728, known for his work in mathematics, physics, and astronomy. He came from a modest family—his father and grandfather were both tailors. Due to financial hardship, Lambert left school at the age of twelve to help his father. Despite this, he was largely self-taught and studied science in his spare time during the evenings.

As a teenager, he held various jobs, including clerk, private tutor, and secretary, and gradually gained recognition in scientific circles. He became a member of important scholarly societies in Switzerland, and in 1755, he published his first scientific paper. In the following years, he released three major books—on the passage of light through different media, on photometry, and on cosmology. In the latter, he proposed the bold idea that the universe is composed of galaxies, which in turn are composed of stars.

In 1764, he was invited to the Prussian Academy of Sciences in Berlin by the renowned Leonhard Euler. Although he sometimes exhibited odd behavior and had an unusual appearance, he eventually gained the trust of King Frederick II. In 1766, Lambert wrote a study on the fifth postulate of Euclidean geometry. By assuming the postulate to be false, he discovered several theorems of non-Euclidean geometry.

M. Ćalasan and S. Vujošević, *Analytical Solutions of Nonlinear Power System Models Using the Lambert W Function*, SpringerBriefs in Electrical and Computer Engineering, https://doi.org/10.1007/978-3-032-09579-4_1

However, in history he would become best known for the so-called Lambert W function. Namely, during the 1750s, Lambert worked on solving a trinomial equation of the form:

$$x^m + px = q. \tag{1.1}$$

Lambert attempted to solve the equation by first determining approximate values of the solution—one from below and one from above. Over time, he noticed that these values gradually converged, ultimately leading to the exact solution of the equation, as follows:

$$x = \frac{q}{p} - \frac{q^m}{p^{m+1}} + m\frac{q^{2m-1}}{p^{2m+1}} - m\frac{3m-1}{2}\frac{q^{3m-2}}{p^{3m+1}}^{\,m} + \dots \tag{1.2}$$

After this, Lambert worked on expanding this equation into a series in order to solve it. Seeing the mathematical problem studied by Lambert, Leonhard Euler extended the considered equation during the 1760s to a much broader class of equations, which have the following form:

$$x^\alpha - x^\beta = (\alpha - \beta)\upsilon x^{\alpha+\beta}. \tag{1.3}$$

Euler considered the limit of both sides of the equation in the case when the value of the variable α approaches the variable β. Namely, in that case, we have the following expression:

$$\lim_{\alpha \to \beta}\left(\frac{x^\alpha - x^\beta}{\alpha - \beta}\right) = \lim_{\alpha \to \beta}\left(\upsilon x^{\alpha+\beta}\right). \tag{1.4}$$

The solution of these equations has the following form:

$$x^\alpha \log(x) = \upsilon x^{2\alpha}, \tag{1.5}$$

that is, the form of the equation

$$\log(x) = \upsilon x^\alpha, \tag{1.6}$$

If both sides of the equation are multiplied by α

$$\log(x^\alpha) = \alpha \upsilon x^\alpha, \tag{1.7}$$

then the previous equation can also be written in the following form,

$$\log(y) = \gamma y, \tag{1.8}$$

where

$$y = x^\alpha \quad \gamma = \alpha \upsilon. \tag{1.9}$$

Finally, if the logarithm operation is replaced with the exponential function, Eq. (1.8) can be written as follows:

$$ze^z = a, \tag{1.10}$$

where

$$z = -\gamma y$$
$$a = -\gamma. \tag{1.11}$$

The solution of this equation is denoted by $W(\alpha)$, and it is known as a solution of Lambert W function.

The explanation of the W notation is described in István Mező's book [1]. In scientific literature, one can find three main reasons why the Lambert function is denoted by the letter W.

- First, in one of the earlier publications, F. N. Fritsch, R. E. Schafer, and W. P. Crowley used the lowercase letter w in the defining equation $W(x)\exp(W(x)) = x$, and this notation was gradually adopted.
- Second, when Gaston H. Gonnet implemented the Lambert function in the Maple mathematical software, he used a capital letter W, since all standard functions in Maple are written in uppercase.
- Third, the letter W also honors the significant contributions of the mathematician E. M. Wright, who conducted in-depth studies of this function between 1949 and 1977.

So, the use of W has mathematical, software-related, and historical justification.

1.2 Basic Properties of the Lambert W Function

The Lambert W function is widely applicable in mathematics, physics, and engineering, especially when trying to find analytical solutions to equations that involve a quantity multiplied by its exponential dependence.

In mathematics, the Lambert W function refers to the various branches that emerge as the inverse of the function defined by [1–4]

$$W(x)e^{W(x)} = x, \tag{1.12}$$

Because this inverse is not unique, the function is represented as multiple branches. The two most notable ones are:

- **The principal branch (W_0)**, which is defined for $x \geq -1/\exp(1)$,
- **The secondary (lower) branch (W_{-1})**, which exists for $-1/\exp(1) \leq x < 0$.

A graphical representation in the real domain typically shows these two branches—$W_0(x)$ in red and $W_{-1}(x)$ in blue (Fig. 1.1). They share a common branch point at $x = -1/\exp(1)$; it is at this point where the function transitions from being single-valued to multi-valued. This multi-branched nature makes the Lambert W function a powerful tool for explicitly representing solutions to a wide range of complex nonlinear problems, which is essential in both technical and natural sciences.

In most engineering and technical applications—such as modeling of solar cells, electrical circuits, diode rectifiers, and similar phenomena—only the principal branch of the Lambert W function, denoted as $W_0(x)$, is used. This is because such applications typically involve solving equations with real and positive solutions. On the other hand, the lower branch W_{-1}, is used in more specific analyses where negative values naturally arise. These include problems in oscillation theory, stability analysis, and the modeling of inductive and capacitive responses.

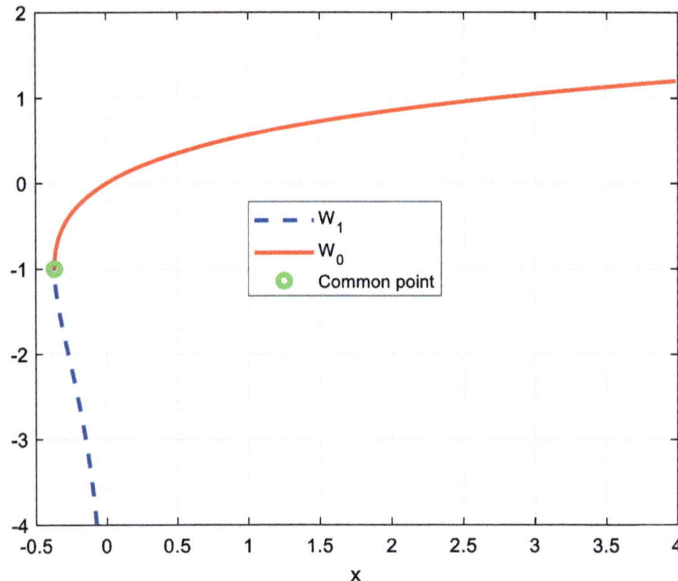

Fig. 1.1 Lambert W function

MATLAB code for plotting the Lambert W equation is given below.

```
clc
clear all
point = 1;
pointW0 = 1;
pointbreak = 1;
for x = -4:0.0001:1.2
    if x<-1
    xW1(point) = x;
    W1(point) = x * exp(x);
    point = point + 1;
    end
    if x==-1
    xbreak(pointbreak) = x;
    Wbreak(pointbreak) = x * exp(x);
    end
    if x>-1
    xW0(pointW0) = x;
    W0(pointW0) = x * exp(x);
    pointW0=pointW0+1;
    end
    end
figure(1)
hold on
plot(W1,xW1,'b--', 'Linewidth', 2)
hold on
plot(W0,xW0,'r', 'Linewidth', 2)
hold on
plot(Wbreak,xbreak,'go', 'Linewidth', 3)
hold on
legend('W_1', 'W_0','Common point')
grid on
box on
xlabel('x')
```

1.3   How Can the Lambert W Equation Be Solved?

The Lambert W function can be computed using both [5, 6]

- analytical, and
- numerical methods.

Among the most common numerical methods are Newton's and Halley's iterations [2, 5, 6], while analytical techniques include Taylor series expansions [6] and the application of the so-called Special Trans Function Theory (STFT) [7–11], which allows accurate solutions in certain domains.

The following sections provide an overview of both numerical and analytical approaches for solving the Lambert W equation, along with a brief discussion about its advantages.

1.3.1   Newton's Method

Newton's method, also known as the Newton–Raphson method, is a numerical algorithm used to find approximate solutions of nonlinear equations of the form $f(x) = 0$. The core idea of the method is to use the tangent line to the function at a given point to iteratively approach a better estimate of the root.

In the case of the Lambert W function, starting from an initial guess x_0, the iterations are computed using the formula [12]:

$$W_1 = W_0 - \frac{W_0 \exp(W_0) - x}{\exp(W_0)(W_0 + 1)}. \tag{1.13}$$

This method is rapidly convergent when the initial guess is close to the actual root and the function is sufficiently smooth. However, if these conditions are not met, it may diverge or converge slowly.

The MATLAB code for solving the Lambert W equation using Newton's method is given below.

```
clc

clear all

W=0;

x=1/5;

tol = 1e-10;

max_iter = 100;

for iter = 1:max_iter
    f = W * exp(W) - x;
```

```
      fp = exp(W) * (1 + W);
      W_new = W - f / fp;
      if abs(W_new - W) < tol
         break;
      end
      W = W_new;

end

W

point=1;

for k=-5:0.0001:0.3
      xaxis(point)=k;
      y1axis(point)=x;
      y2axis(point)=k*exp(k);
      point=point+1;

end

figure(1)

hold on

plot(xaxis,y1axis,'r','Linewidth',2)

hold on

plot(xaxis,y2axis,'b--','Linewidth',2)

legend('x','W*exp(W)')

hold on

hold on

grid on

box on

xlabel('W')
```

1.3.2 Halley's Method

Halley's method is a numerical iterative technique used to find the roots of nonlinear equations, similar to Newton's method but with faster convergence [2]. It is a third-order convergence method, meaning the error decreases more rapidly compared to Newton's second-order method.

The iterative formula is:

$$W_1 = W_0$$
$$- \frac{2(W_0 \exp(W_0) - x)\exp(W_0)(1 + W_0)}{2\left[(\exp(W_0)(1 + W_0))^2\right] - (W_0 \exp(W_0) - x)\exp(W_0)(2 + W_0)}. \quad (1.14)$$

This method requires both the first and second derivatives of the function and is especially useful when high accuracy is needed with fewer iterations.

The MATLAB code for solving the Lambert W equation using Halley's method is given below.

```
clc

clear all

W = 0;

x = 1/5;

tol = 1e-10;

max_iter = 100;

for iter = 1:max_iter
    f = W * exp(W) - x;
    fp = exp(W) * (1 + W);
    fpp = exp(W) * (2 + W);
    W_new = W - (2*f*fp) / (2*fp^2 - f*fpp);
    if abs(W_new - W) < tol
       break;
    end
    W = W_new;

end

W

point = 1;

for k = -5:0.0001:0.3
    xaxis(point) = k;
    y1axis(point) = x;
    y2axis(point) = k * exp(k);
    point = point + 1;

end

figure(1)

hold on

plot(xaxis, y1axis, 'r', 'Linewidth', 2)
```

plot(xaxis, y2axis, 'b--', 'Linewidth', 2)

legend('x', 'W*exp(W)')

grid on

box on

xlabel('W')

1.3.3 Asymptotic Solution

For large values of x, Lambert W equation has the following solution

$$W = L_1 - L_2 + \sum_{l=0}^{\infty}\sum_{m=1}^{\infty} \frac{(-1)^l \begin{bmatrix} l+m \\ l+1 \end{bmatrix}}{m!} L_1^{-l-m} L_2^m, \tag{1.15}$$

where $L_1 = \log(a)$, $L_2 = \log(L_1)$ and $\begin{bmatrix} l+m \\ l+1 \end{bmatrix}$ is a nonnegative Stigling number of the first kind [13].

The advantage of this approach lies in the fact that it relies solely on logarithmic functions and their combinations, making it suitable for analytical processing and integration into models where a simplified representation of system behavior in limiting cases is required. It should be emphasized that the accuracy of the asymptotic expression depends on the number of terms included in the summations. Moreover, this expression is not applicable for small values of x, as it either converges too slowly or is not valid in those regions.

The MATHEMATICA code for solving the Lambert W equation via asymptotic solution is provided below.

```
digitnumber = 300;
x = 5000;
L1 = Log[x];
L2 = Log[L1];
APROX = L1 - L2 + L2/L1 + L2*(-2 + L2)/(2*L1^2) + L2*(6 - 9*L2 + 2*L2^2)/(6*L1^3) +
    L2*(-12 + 36*L2 - 22*L2^2 + 3*L2^3)/(12*L1^4);
Print["solutionAPROX= ", SetPrecision[APROX, digitnumber]];
ErrorAPROX = SetPrecision[Abs[APROX - x*E^(-APROX)], digitnumber];
Print["ErrorAPROX= ", ErrorAPROX]
```

1.3.4 Taylor Series (TS)

The Taylor series is a fundamental tool in mathematical analysis, used to approximate functions by infinite polynomials centered around a specific point. It expresses a function as an infinite sum of its derivatives evaluated at that point, scaled by factorial terms. This approach provides both theoretical insight and practical means for function approximation in physics, engineering, and applied mathematics [6].

The solution of the Lambert W equation using the Taylor series has the following form:

$$W = \sum_{k=1}^{M} \left(\frac{x^k (-k)^{k-1}}{k!} \right).$$

(1.16)

This formula allows for an analytical approximation of the function $W(x)$ without the need for numerical algorithms. However, the number of terms used directly affects the accuracy of the solution—more terms provide more precise results, but increase computational complexity. Additionally, this series represents an approximation of the function for small values of x, specifically for $|x| < 1/\exp(1)$, where the series converges.

Since MATLAB has limitations regarding the maximum value of numbers it can handle, the MATHEMATICA code for solving the Lambert W equation is provided below.

```
digitnumber = 300;
x = 1/6;
M = 50;
                      M
solutionLAMBERT =    ∑  x^k*((-k)^(k-1))/k!;
                     k=1
Print["solutionLAMBERT= ", SetPrecision[solutionLAMBERT, digitnumber]];
ErrorLAMBERT = SetPrecision[Abs[solutionLAMBERT - x*E^(-solutionLAMBERT)], digitnumber];
Print["ErrorLAMBERT= ", ErrorLAMBERT]
```

1.3.5 Special Trans Function Theory

The Special Trans Function Theory (STFT) [7–11] is a modern mathematical approach designed for the approximation of special functions, including the Lambert W function. This method enables solutions to be expressed as rational approximations in the form of finite sums, which significantly simplifies the analytical treatment of complex nonlinear problems. The Lambert W function solution via STFT is as follows:

$$W = x\,\frac{\sum_{k=0}^{M}\left(\frac{x^k(M-k)^k}{k!}\right)}{\sum_{k=0}^{M+1}\left(\frac{x^k(M+1-k)^k}{k!}\right)}. \tag{1.17}$$

Unlike classical series-based methods such as the Taylor expansion, the STFT approach delivers much higher accuracy even for relatively small values of the parameter M, which defines the approximation level. Comparative studies have shown that the Taylor series provides acceptable results only for low values of the parameter x, while its accuracy deteriorates rapidly as x increases. In contrast, STFT maintains consistent precision across the entire range of x, making it a more robust choice.

It is especially worth noting that this method was developed by Prof. Perović from the University of Montenegro and has since found broad application across various engineering disciplines—from nuclear technology and circuit theory to the field of renewable energy, including solar and wind power.

Regardless of the analytical method chosen, the computational error can be estimated using a dedicated formula, while the number of accurate digits in the result—the calculation precision—can also be quantified. Due to its structure and efficiency, STFT represents a significant contribution to the analytical solution of nonlinear equations involving special functions.

The MATHEMATICA code for solving the Lambert W equation via STFT is provided below.

```
digitnumber = 300;

x = 1/6;

M = 50;

        M
F1 =    Σ   (x^k*((M - k)^k)/k!);
       k=0

        M+1
F2 =    Σ   (x^k*((M + 1 - k)^k)/k!);
       k=0

solutionTRANS = x*(F1/F2);

ErrorTRANS = Abs[solutionTRANS - x*E^(-solutionTRANS)];

Print["solutionTRANS= ", SetPrecision[solutionTRANS, digitnumber]];

Print["ErrorTRANS= ", SetPrecision[ErrorTRANS, digitnumber]];
```

Regardless of the analytical approach used, the calculation error (G) can be determined as follows:

$$G = W - x \cdot \exp(-W). \tag{1.18}$$

Based on this error, the precision (P) of the calculation—representing the number of correct digits—can be estimated as follows:

$$P = 1 - \log(G). \tag{1.19}$$

1.3.6 Software Tools

Thanks to its mathematical significance and broad engineering relevance, the Lambert W function has been incorporated into nearly all modern software platforms for numerical and symbolic computation. The notation used for this function varies depending on the environment: in MATLAB, it is denoted as lambertw; in Mathematica, it appears as ProductLog; while Maple uses LambertW. In Python (via the SciPy library), the function is accessed through scipy.special.lambertw, and the GNU Scientific Library provides implementations as gsl_sf_lambert_W0 and gsl_sf_lambert_Wm1. Similar notations are found in other tools such as Octave (lambertw), Maxima (lambert_w), and R (lambertW0, lambertWm1).

1.4 Some Special Values of the Lambert W Function

The Lambert W function, like other special functions, has its own specific properties and characteristic values [1]. Some of these are presented below.

- Properties

$$e^{W(x)} = \frac{x}{W(x)}$$

- Specific value

$$W(0) = 0$$

$$W(e) = 1$$

$$W\left(e \cdot e^e\right) = e$$

$$W\left(-\frac{1}{e}\right) = -1$$

$$W\left(\frac{\sqrt[n]{e}}{n}\right) = \frac{1}{n}$$

- Indefinite integrals

$$\int \frac{W(x)}{x}\,\mathrm{d}x = \frac{(W(x))^2}{2} + W(x) + C$$

- Derivative

$$\frac{\mathrm{d}W(x)}{x} = \frac{W(x)}{x(1 + W(x))} = \frac{1}{x + e^{W(x)}}$$

- Equation of the form $W \cdot b^W = x$

$$W_b(x) = \frac{W(x \cdot \log(b))}{\log(b)}$$

- Equation of the form $W + b^W = x$

$$W = x - \frac{W(b^x \cdot \log(b))}{\log(b)}$$

- Equation of the form $W^W = x$

$$W = \frac{\log(x)}{W(\log(x))}$$

1.5 LogWright or *g*-Function

The implementation of the Lambert W function in software packages depends on the value of its argument. Specifically, for extremely large values, numerical evaluation becomes infeasible due to limitations imposed by the IEEE standard for floating-point representation [14]. If the argument exceeds approximately 10^{323}, which is the upper limit for double-precision numbers, the Lambert W function can no longer be directly applied [15–21].

In such cases, instead of using the Lambert W function, a logarithmic approximation known as the LogWright function, or so-called *g*-functions, is employed. These functions provide a stable way to estimate the result without loss of numerical accuracy. They are particularly useful in symbolic modeling and asymptotic analysis, where very large input values arise that exceed the practical limits of standard numerical algorithms.

The *g*-function has the following form:

$$g = \log(W(x)) = \log(x) - W(x). \tag{1.20}$$

It is evident that the argument of the *g*-function is obtained as a logarithmic transformation of the argument used in the Lambert W function, which significantly simplifies its implementation and enables stable numerical realization.

Figure 1.2 shows the shape of the *g*-function, while the MATLAB code for plotting its graph is provided below. It is evident that for negative argument values, the function exhibits linear behavior, whereas for positive values, it takes on a logarithmic character. The MATLAB script employs Halley's iterative procedure to compute and visualize the *g*-function graph.

```
clc

clear all

counter=0;

for x=-10:0.1:150
        counter=counter+1;
            if x<-exp(1)
               y0=x;
            end
            if x>exp(1)
               y0=log(x);
            end
            if x<exp(1) && x>=-exp(1)
               y0=-exp(1)+(1+exp(1))/(2*exp(1))*(x+exp(1));
            end
            epsilon=10^-15;
            for iter=1:10000
               iteracija(iter)=iter;
               Halleyconv(iter)=y0;
            y1=y0-(2*(y0+exp(y0)-x)*(1+exp(y0)))/(2*(1+exp(y0))^2-(y0+exp(y0)-x)*exp(y0));
                if abs(y1-y0)<epsilon
                   break
                end
                y0=y1;
            end
            gfunction(counter)=(y1);
            xaxis(counter)=x;

end

figure(1)
```

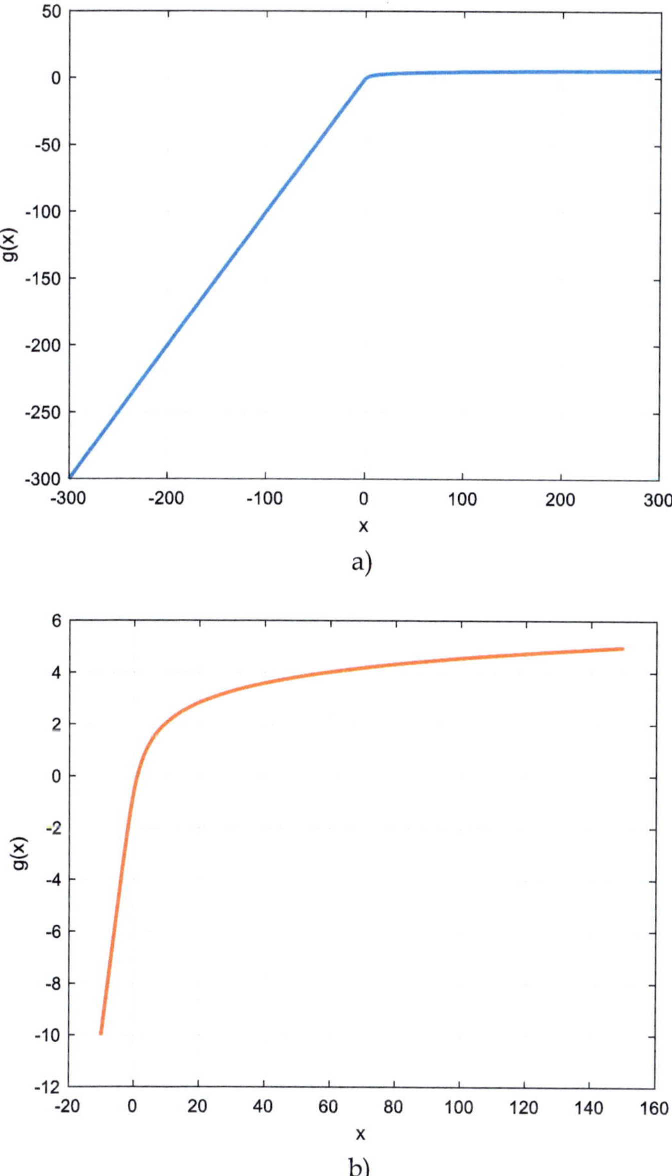

a)

b)

Fig. 1.2 *g*-function **a** full scale, **b** zoomed part

plot(xaxis, gfunction,'Linewidth',2)

hold on

grid on

box on

xlabel('x')

ylabel('g(x)')

1.6   How Can the *g*-Function Be Solved?

The *g*-function, like the Lambert W function, can be computed using both [20]

- **analytical**, and
- **iterative (numerical)** methods.

Any of the well-known numerical methods can be used for iterative solving, and some of them have already been described in the context of solving the Lambert W equation. In this section, two original approaches related to the *g*-function will be presented. First, the so-called *g*-IP, i.e., the *g*-iterative procedure, will be described, followed by the analytical solution based on STFT, referred to as *g*-STFT.

In the literature, it has been shown that Halley's iterative procedure demonstrates the best performance for solving the *g*-function. However, since this method is sensitive to initial conditions, a specific strategy for determining the initial value as a function of the argument has been proposed in various studies. This initialization procedure takes the following form:

$$
\begin{cases}
x \leq -\exp(1) & \to^{(0)} g = x \\
x \geq \exp(1) & \to^{(0)} g = \log(x) \\
-\exp(1) \leq x \leq \exp(1) & \to^{(0)} g = -\exp(1) + \frac{1+\exp(1)}{2\exp(1)}(x + \exp(1))
\end{cases}
\qquad . \qquad (1.21)
$$

The MATHEMATICA code for solving the *g*-function via Halley's method is provided below.

```
Numberofdiggits = 50;
x = 2;
gstartHIP = -E^1 + ((1 + E^1)/(2*E^1))*(x + E^1);
If[x <= -E^1, gstartHIP = x];
If[x >= E^1, gstartHIP = Log[x]];
Print["gstartHIP= ", SetPrecision[gstartHIP, Numberofdiggits]];
criteria = 10^-10;
For[count = 1, count <= 100, count += 1,
     Print["count= ", count];
      gnewHIP =
  gstartHIP - (2*(gstartHIP + E^(gstartHIP) - x)*(1 + E^(gstartHIP)))/
    (2*(1 + E^(gstartHIP))^2 - (gstartHIP + E^(gstartHIP) - x)*E^(gstartHIP));
      Print["HIPdiff= ", SetPrecision[Abs[gstartHIP - gnewHIP], Numberofdiggits]];
      Print["gnewHIP= ", SetPrecision[gnewHIP, Numberofdiggits]];
      If[Abs[gstartHIP - gnewHIP] < criteria, Break[]];
      gstartHIP = gnewHIP;
]
Print["gIP_HIP= ", SetPrecision[gnewHIP, Numberofdiggits]]
```

1.6.1   *g-IP*

The iterative procedure for solving *g*-function has the following form, starting from the initial value of W,

$$W_1 = x - \log(W_0). \tag{1.22}$$

If the stopping criteria is not satisfied the old version of the iteration procedure is calculated as follows

$$W_0 = x - \log(W_1), \tag{1.23}$$

and finally

$$g_1 = \log(W_0). \tag{1.24}$$

The key advantages of the proposed iterative procedure are that it is independent of initial conditions and does not require the computation of derivatives. Moreover, it is well-suited for handling large argument values, making it robust and broadly applicable.

The MATLAB code for solving the *g*-IP procedure is provided below.

clc

clear all

close all

```
x=100;

W0=100;

max_iteration=100;

epsilon=1e-7;

for iteration=1:max_iteration
    W1=x-log(W0);
    if abs(W1-W0)<epsilon
      break
    end
    W0=x-log(W1);

end

gIPsolution=log(real(W0))
```

Also, the MATHEMATICA code for solving the g-function via g-IP is provided below.

```
Numberofdiggits = 50;
x = 2;
W0 = 1;
criteria = 10^-10;
For [count = 1, count <= 20, count += 1,
        Print ["count= ", count];
        W1 = x - Log[W0];
        Print ["gIP=", SetPrecision[Abs[xstart - xnew], Numberofdiggits]];
        Print["gIP= ", SetPrecision[Log[xnew], Numberofdiggits]] ;
        If[Abs[W1 - W0] < criteria, Break[]];
        W0 = x - Log[W1];
]
gIP = Log[W1];
Print["gIP= ", SetPrecision[gIP, Numberofdiggits]]
```

1.6.2 g-STFT

For the analytical evaluation of the g-function, a direct application of the STFT approach can also be used. Namely, knowing that the argument of the g-function corresponds to the logarithmic value of the argument in the Lambert W function, and that the g-function itself represents the logarithmic value of the Lambert W function, the following expression can be written:

$$g = x + \log\left(\frac{\sum_{n=0}^{M} \frac{e^{x \cdot n}(M-n)^n}{n!}}{\sum_{n=0}^{M+1} \frac{e^{x \cdot n}(M+1-n)^n}{n!}}\right). \tag{1.25}$$

As can be seen from the equation itself, the sum contains an exponential value of the g-function argument. As a result, this analytical solution is very efficient for small argument values, while numerical problems may arise for large argument values.

The MATHEMATICA code for solving the g-function via g-STFT is provided below.

```
Numberofdiggits = 50;
x = 2;
Print["x =", x]
M = 100;
         M
F1  =   ∑  (E^(x*k)*(((M - k)^k)/k!));
        k=0

         M+1
F2  =   ∑  (E^(x*k)*(((M + 1 - k)^k)/k!));
        k=0

GSTFT = SetPrecision[x + Log[(F1/F2)], Numberofdiggits];
Print["G_function = ", GSTFT]
GSTFTCalculationError = SetPrecision[Abs[GSTFT - x + E^GSTFT], Numberofdiggits];
Print["Calculation_error= ", GSTFTCalculationError]
```

1.7 Numerical Examples

This section presents some of the numerical results. First, a comparison between the STFT and TS methods was performed for different values of the parameter M and various values of the Lambert W function argument. These results are presented both in tabular (see Table 1.1) and graphical form (see Fig. 1.3).

It can be observed that the STFT method generally yields more accurate or consistent values of P compared to the TS method, especially for smaller values of the parameter x. This highlights the robustness of the STFT approach in handling different ranges of the input parameters.

Table 1.2 presents the numerical results for solving the g-function using both Halley's method and the proposed g-IP procedure. The corresponding results illustrating the influence of accuracy and initial conditions on the g-IP method are also shown in Fig. 1.4. As observed, for lower values of the argument x, Halley's method demonstrates faster convergence. However, as x increases, the difference between the two methods becomes negligible. Importantly, the g-IP method consistently converges regardless of the initial conditions, confirming its robustness and independence from the initial guess.

Finally, in Table 1.3, the numerical results of the accuracy of the g-STFT method are presented for various values of the argument of the g-function. As can be seen, for negative and small values of the g-function argument, it is sufficient to use only a few terms in the sum (i.e., a small M). However, for large values of the g-function argument, larger values of M are required.

Table 1.1 Computed values of the parameter P using STFT and TS methods for different values of M and x

x	M	W	G_{STFT}	P_{STFT}	G_{TS}	P_{TS}
0.0001	10	0.0000999900014997333	$2.41\cdot10^{-56}$	57	$6.49\cdot10^{-42}$	43
	20	85405869000452213…	$3.17\cdot10^{-107}$	108	$5.44\cdot10^{-78}$	79
	30		$3.78\cdot10^{-158}$	159	$6.69\cdot10^{-114}$	115
	40		$3.96\cdot10^{-209}$	210	$9.70\cdot10^{-150}$	151
	50		$3.29\cdot10^{-260}$	261	$1.54\cdot10^{-185}$	186
0.001	10	0.0009990014973385308	$1.66\cdot10^{-44}$	45	$6.48\cdot10^{-31}$	32
	20	8995782787410778…	$4.95\cdot10^{-85}$	86	$5.43\cdot10^{-57}$	58
	30		$1.17\cdot10^{-124}$	125	$6.68\cdot10^{-83}$	84
	40		$3.06\cdot10^{-164}$	165	$9.68\cdot10^{-109}$	110
	50		$4.66\cdot10^{-204}$	205	$1.53\cdot10^{-134}$	135
0.02	10	0.0196115893374056272	$2.23\cdot10^{-28}$	29	$1.29\cdot10^{-16}$	17
	20	9168248268298370…	$6.07\cdot10^{-54}$	55	$1.10\cdot10^{-29}$	30
	30		$7.91\cdot10^{-80}$	81	$1.39\cdot10^{-42}$	43
	40		$2.48\cdot10^{-105}$	106	$2.06\cdot10^{-55}$	56
	50		$2.21\cdot10^{-130}$	131	$2.21\cdot10^{-68}$	69
0.10	10	0.0912765271608622643	$4.01\cdot10^{-20}$	21	$5.72\cdot10^{-9}$	10
	20	36684101958239000…	$5.94\cdot10^{-38}$	39	$4.74\cdot10^{-15}$	16
	30		$4.43\cdot10^{-56}$	57	$5.80\cdot10^{-21}$	22
	40		$1.15\cdot10^{-72}$	73	$8.38\cdot10^{-27}$	28
	50		$3.04\cdot10^{-92}$	93	$1.33\cdot10^{-32}$	33
0.15	10	0.1315149280010344580	$5.81\cdot10^{-18}$	19	$4.68\cdot10^{-7}$	8
	20	4744166673662417…	$2.32\cdot10^{-34}$	35	$2.22\cdot10^{-11}$	12
	30		$1.85\cdot10^{-50}$	51	$1.56\cdot10^{-15}$	16
	40		$3.04\cdot10^{-66}$	67	$1.30\cdot10^{-19}$	20
	50		$1.30\cdot10^{-82}$	83	$1.19\cdot10^{-23}$	24
0.30	10	0.2367553107885593168	$8.36\cdot10^{-16}$	17	$8.31\cdot10^{-4}$	5
	20	7136699131310225…	$1.16\cdot10^{-28}$	29	$4.00\cdot10^{-5}$	6
	30		$1.60\cdot10^{-41}$	42	$2.87\cdot10^{-6}$	7
	40		$1.40\cdot10^{-54}$	55	$2.45\cdot10^{-7}$	8
	50		$1.02\cdot10^{-67}$	68	$2.29\cdot10^{-8}$	9

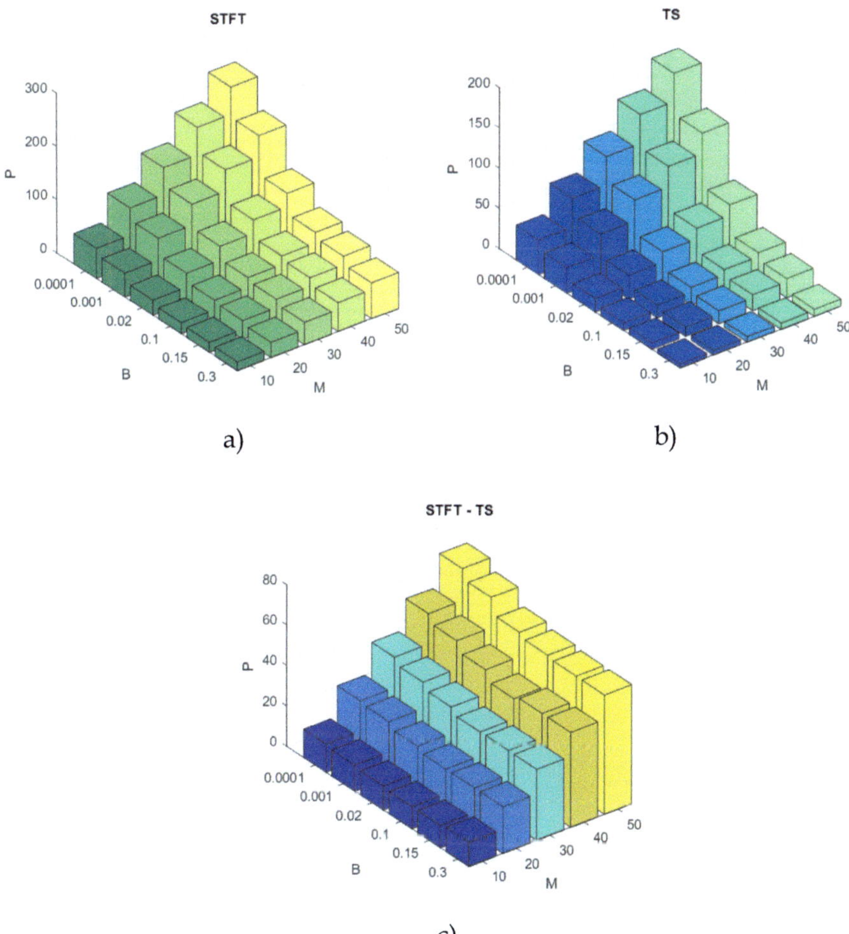

Fig. 1.3 Numerical results for **a** STFT, **b** TS and **c** difference between STFT and TS

Table 1.2 Numerical results for solving the g-function via Halley's method and g-IP

x	g	Stopping criterion	Required number of iterations for Halley's method	Initial conditions for g-IP - W_0	Required number of iterations for g-IP
5	1.3065586410	$< 10^{-5}$	3	1/1000	7
				1	7
				100	7
		$< 10^{-10}$	3	1/1000	12
				1	11
				100	12
50	3.8322808345	$< 10^{-5}$	3	1/1000	4
				1	4
				100	3
		$< 10^{-10}$	3	1/1000	5
				1	5
				100	5
100	4.5585133544	$< 10^{-5}$	2	1/1000	4
				1	3
				100	3
		$< 10^{-10}$	3	1/1000	5
				1	5
				100	4
500	6.2021262709	$< 10^{-5}$	3	1/1000	3
				1	3
				100	3
		$< 10^{-10}$	3	1/1000	4
				1	4
				100	4
1000	6.9008305276	$< 10^{-5}$	2	1/1000	3

(continued)

Table 1.2 (continued)

x	g	Stopping criterion	Required number of iterations for Halley's method	Initial conditions for g-IP - W_0	Required number of iterations for g-IP
				1	3
				100	3
		$< 10^{-10}$	3	1/1000	4
				1	4
				100	4
5000	8.5154886417	$< 10^{-5}$	3	1/1000	3
				1	3
				100	3
		$< 10^{-10}$	3	1/1000	4
				1	3
				100	3
50,000	10.819561869	$< 10^{-5}$	2	1/1000	3
				1	3
				100	3
		$< 10^{-10}$	2	1/1000	3
				1	3
				100	3
100,000	11.512810330	$< 10^{-5}$	2	1/1000	3
				1	3
				100	3
		$< 10^{-10}$	2	1/1000	3
				1	3
				100	3

Fig. 1.4 Number of Iterations for g-IP Method for different value of criteria **a** 10^{-5} and **b** 10^{-10}

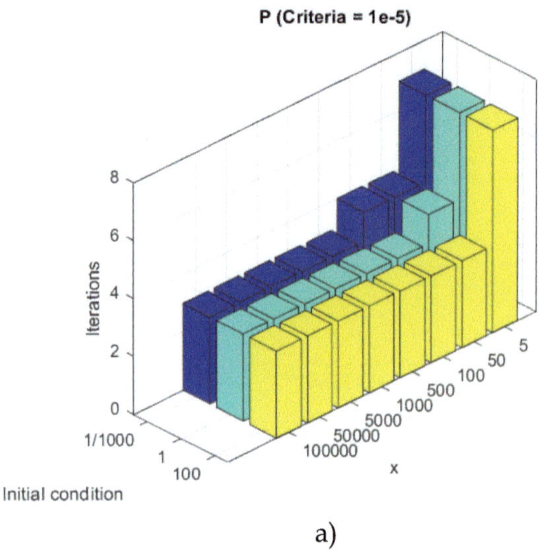

a)

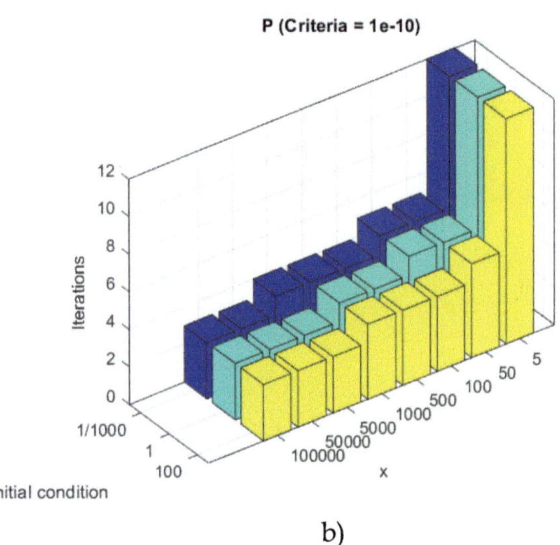

b)

Table 1.3 Numerical results for the g-STFT

x	Parameter M	g-function value	G	P
-50	2	$-50.000000000000000000000192\ldots$	$1.1958 \cdot 10^{-66}$	67
-10	2	$-10.0000453978687\ldots$	$1.5593 \cdot 10^{-14}$	15
	3	$-10.00004539786874\ldots$	$1.7695 \cdot 10^{-19}$	20

(continued)

Table 1.3 (continued)

x	Parameter M	g-function value	G	P
-5	2	$-5.00669\ldots$	$4.9716\cdot10^{-8}$	9
	3	$-5.006693000\ldots$	$8.1824\cdot10^{-11}$	12
	4	$-5.006693000497\ldots$	$1.0635\cdot10^{-13}$	14
	5	$-5.0066930004977311\ldots$	$1.1243\cdot10^{-16}$	17
-3	3	$-3.047478\ldots$	$1.7995\cdot10^{-7}$	8
	4	$-3.04747848\ldots$	$1.3415\cdot10^{-9}$	10
	5	$-3.0474784910\ldots$	$5.9595\cdot10^{-12}$	13
	6	$-3.04747849102\ldots$	$8.3294\cdot10^{-15}$	16
-1	4	$-1.27846\ldots$	$4.5727\cdot10^{-6}$	7
	5	$-1.278464\ldots$	$4.5132\cdot10^{-7}$	8
	7	$-1.27846454\ldots$	$6.9419\cdot10^{-10}$	11
	9	$-1.27846454276\ldots$	$2.7421\cdot10^{-12}$	13
	15	$-1.278464542761\ldots$	$1.1704\cdot10^{-19}$	20
$1/2$	7	$-0.266248\ldots$	$7.5032\cdot10^{-7}$	8
	12	$-0.266248608\ldots$	$3.6813\cdot10^{-10}$	11
	14	$-0.2662486081\ldots$	$2.5815\cdot10^{-12}$	13
	18	$-0.26624860816175\ldots$	$9.6548\cdot10^{-15}$	16
	23	$-0.266248608161750\ldots$	$5.5089\cdot10^{-19}$	20
2	10	$0.4428\ldots$	$3.2985\cdot10^{-6}$	7
	15	$0.442854\ldots$	$4.2002\cdot10^{-8}$	9
	20	$0.442854401\ldots$	$2.8147\cdot10^{-10}$	11
	30	$0.442854401002388\ldots$	$6.5533\cdot10^{-16}$	17
	40	$0.44285440100238858\ldots$	$5.3386\cdot10^{-20}$	21
5	20	$1.3065\ldots$	$7.5057\cdot10^{-5}$	6
	35	$1.30655864\ldots$	$4.6877\cdot10^{-9}$	10
	50	$1.3065586410\ldots$	$1.3097\cdot10^{-11}$	12
	80	$1.30655864103\ldots$	$8.9317\cdot10^{-18}$	19
10	100	$2.0705799\ldots$	$5.2465\cdot10^{-8}$	9
	150	$2.070579904980\ldots$	$4.0668\cdot10^{-12}$	13
	300	$2.07057990498030\ldots$	$1.7286\cdot10^{-24}$	25
50	2000	$3.8322808\ldots$	$3.9371\cdot10^{-7}$	8
	5000	$3.83228083\ldots$	$6.4152\cdot10^{-20}$	21

References

1. Mező I (2022) The Lambert W function: its generalizations and applications. Chapman and Hall/CRC (Discrete mathematics and its applications). https://doi.org/10.1201/978100316 8102. ISBN 978-0367766832
2. Veberic D (2012) Lambert W function for applications in physics. Comput Phys Commun 183(12):2622–2628. https://doi.org/10.1016/j.cpc.2012.07.008
3. Corless RM, Gonnet GH, Hare DEG, Jeffrey DJ, Knuth DE (1996) On the Lambert W function. Adv Comput Math 5:329–359. https://doi.org/10.1007/BF02124750
4. Lankireddy P, Jeevanandam S, Chaudhary A, Deshmukh PC, Roberts K, Valluri SR (2023) Solar cells, Lambert W and the LogWright functions. https://doi.org/10.48550/arXiv.2307.08099
5. Calasan M (2019) Analytical solution for no-load induction machine speed calculation during direct start-up. Int Trans Electr Energy Syst 29(4):e2777. https://doi.org/10.1002/etep.2777
6. Ćalasan M, Abdel Aleem SHE, Zobaa AF (2020) On the root mean square error (RMSE) calculation for parameter estimation of photovoltaic models: a novel exact analytical solution based on Lambert W function. Energy Convers Manag 210:112716. https://doi.org/10.1016/j.enconman.2020.112716
7. Perovich SM, Djukanovic M, Dlabac T, Nikolic D, Calasan M (2015) Concerning a novel mathematical approach to the solar cell junction ideality factor estimation. Appl Math Model 39(12):3248–3264. https://doi.org/10.1016/j.apm.2014.11.026
8. Perovich SM, Orlandic M, Calasan M (2013) Concerning exact analytical STFT solution to some families of inverse problems in engineering material theory. Appl Math Modell 37(7):5474–5497. https://doi.org/10.1016/j.apm.2012.10.052
9. Perovich SM, Calasan M, Toskovic R (2014) On the exact analytical solution of some families of equilibrium critical thickness transcendental equations. AIP Adv 4(11):117124–1171132. https://doi.org/10.1063/1.4902161
10. Perovich SM (1992) Transcendental method in the theory of neutron slowing down. J Phys A Math Gen 25:2969–2989. https://doi.org/10.1088/0305-4470/25/10/024
11. Perovich SM (1997) Concerning the analytical solution of the disperse equation in the linear transport theory. Transp Theory Stat Phys 6:705. https://doi.org/10.1080/00411459708229331
12. Ćalasan M, Al-Dhaifallah M, Ali ZM, Abdel Aleem SHE (2022) Comparative analysis of different iterative methods for solving current-voltage characteristics of double and triple diode models of solar cells. Mathematics 10:3082. https://doi.org/10.3390/math10173082
13. Ćalasan M, Nedić A (2018) Experimental testing and analytical solution by means of Lambert W-function of inductor air gap length. Electr Power Compon Syst 46(7):852–862. https://doi.org/10.1080/15325008.2018.1488012
14. IEEE Standard for Floating-Point Arithmetic, IEEE Std 754™2008, IEEE Computer Society Sponsored by the Microprocessor Standards Committee (2008)
15. Roberts K (2015) A robust approximation to a lambert-type function. arXiv preprint arXiv:1504.01964
16. Roberts K, Valluri SR (2015) On calculating the current-voltage characteristic of multidiode models for organic solar cells. arXiv preprint arXiv:1601.02679
17. Corless RM, Jeffrey DJ (2002) The wright ω function. In: International conference on artificial intelligence and symbolic computation. Springer, pp 76–89
18. Toledo FJ, Herranz MV, Blanes JM, Galiano V (2022) Quick and accurate strategy for calculating the solutions of the photovoltaic single diode model equation. IEEE J Photovolt 12(2):493–500
19. Ćalasan M (2025) Diode-resistance circuit—Novel exact and approximate solutions based on the g-function approach. AEU Int J Electron Commun 193:155744. https://doi.org/10.1016/j.aeue.2025.155744

20. Calasan M (2025) Iterative methods for solving g-functions: a review, comparative evaluation, and application in the solar cell domain. J Comput Electron 24(58) (2025). https://doi.org/10.1007/s10825-025-02298-2
21. Ćalasan M (2025) Double-diode and triple-diode solar cell models: invertible approximate analytical expressions based on the g-function approach. J Comput Electron 24:18. https://doi.org/10.1007/s10825-024-02259-1

Chapter 2
Application of Lambert W Function in Solar Cell Modeling

This chapter is divided into three parts. The first part provides fundamental information on solar energy and photovoltaic cells. The second part discusses the application of the Lambert W function and the g-function in modeling the current–voltage and voltage–current characteristics of solar cells. The third part presents a numerical example illustrating the implementation of the proposed approach.

2.1 Basic Information on Solar Energy and Photovoltaic Cells

Modern society finds itself at a pivotal point in the development of the energy sector. On one hand, there is a growing need to preserve the environment and reduce dependence on fossil fuels; on the other hand, rapid technological advancement is enabling the realization of the sustainable development concept, which until recently existed only at a theoretical level. In this context, renewable energy sources are no longer viewed as alternative pathways but are becoming integral components of long-term energy policies and strategies.

The global transition of the energy system increasingly relies on boosting the share of renewable energy in the overall energy mix [1]. This shift not only contributes to reducing carbon dioxide emissions and protecting the environment but also addresses rising demands for energy security, source diversification, and system resilience.

Within this modern framework, renewable energy sources, thanks to their natural availability, environmental compatibility, and continuous technological progress, occupy a central role in the stable and sustainable supply of end users. Solar, wind, and hydro energy are particularly prominent, as they represent the most widely deployed forms of renewable potential in both technical and market applications.

Among the available renewable resources, solar energy stands out due to its universal presence, geographic accessibility, and ability to be directly converted

M. Ćalasan and S. Vujošević, *Analytical Solutions of Nonlinear Power System Models Using the Lambert W Function*, SpringerBriefs in Electrical and Computer Engineering, https://doi.org/10.1007/978-3-032-09579-4_2

into electricity through photovoltaic (PV) cells. Unlike other sources that require mechanical converters or intermediate stages of energy transformation, solar cells enable direct conversion from radiation to electricity without moving parts, noise, or intensive maintenance.

Thanks to these advantages, photovoltaic systems are increasingly being applied in various domains—from individual households and rural microsystems to large-scale solar power plants connected to the electrical grid. Nevertheless, despite their structural simplicity, the operation of PV cells involves complex physical processes, whose electrical output characteristics depend on irradiance, temperature, material properties, and operating conditions.

Due to this complexity, the development of reliable mathematical models for solar cells is fundamental for simulating, analyzing, and optimizing solar systems. Such models enable accurate prediction of the cell's output behavior under various operating modes, as well as proper system sizing concerning location, load, and target energy efficiency.

To that end, the electrical output of a solar cell can be most comprehensively described through its current–voltage (I–V) characteristic. This relationship provides essential insights into the cell's operation under different load conditions, including the maximum power point, open-circuit voltage, and short-circuit current. The challenge in modeling this relationship lies in its highly nonlinear nature, which arises from complex interactions of photogenerated carriers, reverse current flow through the diode structure, and the presence of parasitic parameters such as series and shunt resistances [2].

Depending on the analysis objectives, required accuracy, and available data, various modeling schemes are applied in practice. The most commonly used are single-diode and two-diode models, although additional elements may be introduced in some cases to better approximate cell behavior under extreme conditions. Each of these configurations strikes a balance between mathematical complexity and physical realism. Regardless of the chosen structure, the reliability of the model depends on accurate parameter identification: saturation current, photogenerated current, ideality factor, and the values of the series and parallel resistances.

Irrespective of the specific solar cell model structure used, three main approaches are typically employed for parameter estimation [3]:

- analytical expressions,
- numerical methods, and
- metaheuristic optimization methods.

Analytical techniques rely on deriving closed-form mathematical expressions that link known points on the I–V curve with the model's unknown parameters. These approaches are often based on manufacturer datasheets or a limited number of measured values. However, due to the need to simplify expressions, various approximations are introduced in practice, which may compromise accuracy.

In contrast, numerical methods, based on iterative algorithms, allow for greater precision as they take into account the full nonlinear form of the model. However, their efficiency depends on the quality of initial assumptions, step size selection, and the stability of the numerical procedure itself. For this reason, engineering practice often combines analytical expressions with numerical techniques to strike a balance between simplicity and accuracy.

In recent years, metaheuristic algorithms have gained considerable traction for solar cell parameter optimization. These methods have proven to be particularly effective for solving parameter identification problems, especially when models are highly nonlinear and the available experimental data is limited or affected by measurement noise. Their advantages include the ability to find globally optimal solutions without requiring knowledge of the objective function's structure or its derivatives, making them highly flexible for different modeling contexts. Additionally, these algorithms exhibit strong robustness and resistance to local minima, which is particularly important in optimizing complex functional relationships that characterize solar cells. Owing to their flexibility, resilience to local optima, and ability to handle nonlinear, multidimensional objective functions, these algorithms have become powerful tools for parameter estimation under realistic operating conditions, even when the experimental database is limited.

When it comes to equivalent circuits used for modeling solar cells, various techniques are applied that differ in both structural complexity and achievable accuracy. These models aim to represent the real-world electrical behavior of the cell using a set of idealized electronic components, thereby facilitating the analysis, simulation, and optimization of solar systems.

The most basic and widely adopted approach is the Single Diode Model (SDM), which captures the core mechanisms of energy conversion in PV cells with relatively low computational complexity [4]. This model consists of a photogenerated current source, an ideal PN junction diode, a series resistance R_S accounting for internal and contact resistances, and a shunt resistance R_P modeling leakage currents and material imperfections. With this structure, the SDM achieves a good balance between physical interpretability and mathematical tractability, making it suitable for a broad range of applications—from engineering design to numerical optimization. For this reason, the SDM has become a reference point in most contemporary approaches to solar cell modeling.

Moreover, thanks to its many advantages—particularly its analytical simplicity and the possibility of deriving closed-form I–V expressions—the single-diode model retains an important role in modern modeling methodologies and is still widely used in both technical and research practices. In more recent literature, to enhance accuracy and adaptability to real-world data, various modified versions of the SDM have been developed. These variants introduce certain structural and mathematical adjustments to improve agreement with experimental I–V curves across a broader range of operating conditions, thereby enabling more precise representation of the nonlinear effects in solar cell performance.

Another frequently used model in technical and research literature is the Double Diode Model (DDM), which extends the basic SDM approach by incorporating a second diode [3, 4]. This additional diode enables a more precise description of carrier recombination mechanisms in both the bulk and depletion regions of the PN junction—particularly relevant under low illumination or for high-efficiency cells.

Compared to the SDM, the DDM provides better alignment with experimental I–V data but at the cost of increased model complexity and a greater number of parameters requiring identification.

Further structural extensions of diode-based models have led to more advanced concepts, such as the Triple Diode Model (TDM) [3, 4] and the Four Diode Model (FDM) [5, 6]. These models aim to encompass an even broader spectrum of physical phenomena affecting solar cell operation, including different recombination mechanisms, material degradation effects, and nonlinear behavior under extreme operating conditions. Each additional diode in the model allows for more accurate isolation of individual transport process contributions, but simultaneously increases the number of parameters—adding complexity both in terms of mathematical formulation and experimental identification.

2.2 Solar Cell Equivalent Circuits

This section describes the SDM, DDM, TDM, and FDM equivalent circuits of solar cells, highlighting the application of the Lambert W function.

2.2.1 SDM

The Single Diode Model (SDM) is the most commonly used equivalent circuit for the analytical and numerical description of solar cell behavior. Its structure is based on the fundamental physical processes occurring within the photovoltaic junction and represents a compromise between simplicity and sufficient accuracy for a wide range of applications.

The standard configuration of the SDM includes four main elements:

- the photo-generated current source I_{PV}, which models the current generation due to solar radiation absorption;
- an ideal diode with an ideality factor a, describing the nonlinear current flow associated with recombination processes within the junction;
- a series resistance R_S, accounting for losses in contacts, material, and metallization;
- a parallel (shunt) resistance R_P, simulating leakage currents caused by structural defects and other non-idealities.

Figure 2.1 shows the electrical schematic of the Single Diode Model (SDM), which includes a photo-generated current source, an ideal diode, as well as series and parallel resistances. This configuration enables the key mechanisms within the solar cell to be represented by a simple yet sufficiently representative electrical circuit.

It is important to emphasize that the arrangement of elements in the SDM is not arbitrary but is grounded in clear physical principles:

Fig. 2.1 SDM

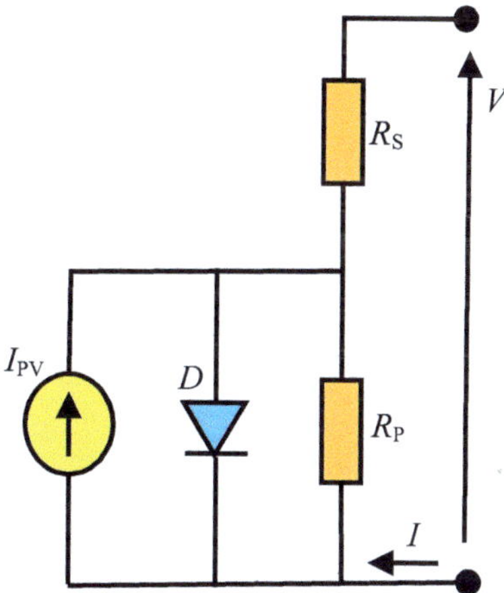

- The diode is placed in parallel with the current source because it models the recombination of charge carriers occurring within the bulk of the semiconductor—a process that happens concurrently with photo-generation.
- The series resistor R_S is connected in series with the external circuit, as it accounts for all losses encountered by the current along its path from photo-generation to the output terminals—including metal contacts, junctions, and material resistances.
- The parallel resistor R_P is also connected in parallel, as it simulates leakage currents through poorly insulated parts of the structure, thereby representing an alternative current path besides the output terminals.

As previously mentioned, the electrical output of a solar cell is typically represented by its current-voltage (I-V) characteristic, which provides insight into the cell's behavior under various load conditions. Within the SDM framework, this dependency is described by a nonlinear equation relating the cell's output current and voltage, taking into account physical and technological parameters such as saturation current, photo-generated current, series and parallel resistances, and the diode ideality factor.

Mathematically, the behavior of a solar cell in the SDM is described using a nonlinear algebraic relationship between the output current and voltage. This relationship fully encompasses the fundamental physical mechanisms occurring in the cell's structure—the generation of current due to solar irradiation, recombination of carriers in the semiconductor junction, energy dissipation through internal resistances, and current leakage through parasitic paths.

As a result of these physical processes, the current-voltage characteristic of a solar cell within the SDM can be formally expressed by the following nonlinear equation:

$$I = I_{PV} - I_0\left(e^{\frac{V+IR_S}{aV_{th}}} - 1\right) - \frac{V + IR_S}{R_P},\tag{2.1}$$

where

I—output current of the solar cell [A],

V—output voltage [V],

I_{PV}—photo-generated current, proportional to the light intensity [A],

I_0—diode saturation current [A],

a—diode ideality factor (typically between 1 and 2),

R_S—series resistance [Ω],

R_P—shunt (parallel) resistance [Ω],

V_{th}—thermal voltage ($=kBT/q$) [V],

q—elementary charge ($\approx 1.602 \times 10^{-19}$ C),

kB—Boltzmann constant ($\approx 1.381 \times 10^{-23}$ J/K),

T—absolute temperature [K].

It is important to highlight that the nonlinear current-voltage relationship in the SDM, although seemingly unsolvable for the output current in closed form, can be analytically expressed using the Lambert W function. This allows for an explicit closed-form solution for the current I, eliminating the need for numerical iteration.

The analytical solution can then be written in the following form:

$$I = \frac{R_P(I_{PV} + I_0) - V}{R_S + R_P} - \frac{aV_{th}}{R_S} W\left(\frac{I_0 R_S R_P}{aV_{th}(R_S + R_P)} e^{\frac{R_P(R_S I_{PV} + R_S I_0 + V)}{a(R_P + R_S)}} \right). \tag{2.2}$$

In the above expression, the symbol W denotes the solution of the Lambert W function. It is well known that the current–voltage (I–V) characteristic of a solar cell represents a unique functional dependence—each voltage value corresponds to a single current value. Therefore, this formulation enables the direct computation of output current for a given voltage, which is particularly useful for comparison with experimental results and model validation. This closed-form representation is especially advantageous in applications requiring repeated evaluation of the I–V curve, such as optimization routines, maximum power point tracking (MPPT) algorithms, or when integrating the model into larger simulation environments.

Although most solar cell models primarily analyze current as a function of voltage, there are numerous practical scenarios in which expressing voltage as a function of current is equally important—or even necessary. For instance, when comparing model results with experimental I–V data, measurements are often performed at predefined current levels, with voltage being the measured outcome. Moreover, in optimization algorithms involving parameter identification, it is often more convenient to use voltage as the dependent variable, since minimizing errors (e.g., RMSE or MAE) aligns more directly with actual measured conditions.

In addition, expressing voltage as a function of current allows direct application in simulations where current is known as an input (e.g., in cascaded current-driven circuits or during MPPT algorithm modeling), and the model needs to predict the corresponding voltage.

While voltage as a function of current can be expressed analytically, this expression is not universally applicable for all current values. In particular, limitations arise under no-load conditions and at very low current levels, where the approach becomes numerically infeasible. The reason lies in the fact that the argument of the Lambert W function can reach extremely high values—often exceeding 10^{323}, which surpasses the computational limits of most software environments (based on IEEE standard 754-2008). This situation is especially pronounced in solar cells with a high shunt resistance (e.g., greater than 1000 Ω) [7–9].

As an alternative, a stable and accurate analytical relationship between voltage and current—valid across the entire operating range—can be established by using the g-function, also known as the LogWright function. This approach enables reliable modeling even in the boundary operating regimes of the solar cell. By applying the g-function, along with appropriate mathematical transformations, it becomes possible to obtain an analytical expression for voltage as a function of current. This allows for precise resolution of the model even in conditions where standard Lambert W-based approaches become numerically unstable or inapplicable.

In this case, the analytical expression for voltage V is given by the following relation [10]:

$$V = aV_{\text{th}}\left(g(\alpha_{\text{V}}) - \log\left(\frac{I_0 R_{\text{P}}}{aV_{\text{th}}}\right)\right) - R_{\text{S}}I, \tag{2.3}$$

where

$$\alpha_{\text{V}} = \log\left(\frac{I_0 R_P}{aV_{\text{th}}}\right) + \frac{R_{\text{P}}(I_{\text{PV}} + I_0 - I)}{aV_{\text{th}}}. \tag{2.4}$$

As previously stated, such a formulation is particularly useful when assessing model accuracy based on experimental data, where measurements are often made for known current values, and voltage is the dependent variable. In this context, having a direct voltage–current relationship allows for the application of error metrics such as RMSE, MAE, and MAPE without requiring additional numerical inversion of the model. This significantly improves both the efficiency and accuracy of the solar cell parameter identification process, especially in cases involving pronounced nonlinearities and high-resolution measurements.

Beyond the standard model, numerous enhanced versions of the single-diode model (SDM) have been developed in the literature to improve modeling precision. Some of these modifications involve rearrangements of model components, aiming to simplify analytical formulation or improve numerical stability. Representative strategies include [11–16]:

- Transposition of model components, i.e., rearranging the position of the diode and resistive elements to simplify the analytical formulation or enhance numerical stability.

- Insertion of an additional resistive element in series with the diode, used to simulate internal losses within the PN junction, improves accuracy in the mid-voltage operating region.
- Addition of an extra shunt resistor, commonly denoted as R_D, introduced as an empirical correction to account for specific operational regimes of the solar cell.

A tabular overview of various modified SDM models of solar cells is presented below (Table 2.1), including the equivalent circuit, the mathematical formulation of the current–voltage relationship, and corresponding MATLAB code lines defining these dependencies.

In addition to these structural modifications, there are also enhanced SDM variants incorporating nonlinear parameter dependencies to better reflect real-world solar cell behavior under varying operating conditions [17, 18]. Examples include:

- Modeling resistances as voltage-dependent functions, wherein the series resistance R_S and shunt resistance R_P are no longer constants but adapt to the operating point, especially in high- or low-voltage regions.
- Dynamic ideality factor, where the diode ideality factor a becomes voltage-dependent, allowing for a more accurate representation of recombination processes occurring in different regions of the solar cell.

2.2.2 DDM

As part of advanced approaches to solar cell modeling, the Double Diode Model (DDM) represents a significant enhancement compared to the classical single-diode structure [19]. The primary motivation behind the development of this model lies in the need for a more accurate description of the complex recombination processes occurring both in the bulk region of the PN junction and in its depletion region—phenomena that cannot be adequately represented by simpler models.

Unlike the Single Diode Model (SDM), which uses a single ideal diode to model recombination, the DDM introduces an additional diode with a separate ideality factor. This allows for improved alignment with experimental I–V data, particularly in low-irradiance conditions and for high-efficiency cells. As a result, simulation accuracy is significantly improved, especially in the regions of the curve where the SDM exhibits deviations—for example, during the transition from the linear to the exponential regime.

This improved accuracy, however, comes at the cost of increased complexity: the number of parameters that must be identified is higher, and the optimization procedures become more demanding. This requires more robust methods and higher-quality experimental data.

Table 2.1 SEM equivalent circuit models

Equivalent circuit	Equation and MATLAB code

	$$I = \frac{R_P(I_{PV} + I_0) - V}{R_S + R_P} - \frac{aV_{th}}{R_S}W(x)$$ $$x = \frac{I_0 R_S R_P}{aV_{th}(R_S + R_P)}e^{\frac{R_P(R_S I_{PV} + R_S I_0 + V)}{a(R_P + R_S)}}$$ **Matlab code**:
	 `x=I0*Rs*Rp/(a*Vth*(Rs+Rp))*exp((Rp*(Rs*Ipv+Rs*I0+V))/(a*Vth*(Rs+Rp)));`
	 `I=(Rp*(Ipv+I0)-V)/(Rs+Rp)-a*Vth*lambertw(x)/Rs;`

	$$I = \frac{R_P}{R_S+R_P}\left(I_{PV} + I_{01} - \frac{V}{R_P} - W(x_{ISDM})\right)$$ $$x_{ISDM} = \frac{R_P R_S + R_D(R_S + R_P)}{R_P + R_S}\frac{I_{01}}{aV_{th}}e^{\frac{1}{aV_{th}}\left(V\left(1+\frac{R_{SD}}{R_P}\right) - R_{SD}I_{PV} + \frac{R_P R_S + R_D(R_S + R_P)}{R_P + R_S}\left(I_{PV} + I_{01} - \frac{V}{R_P}\right)\right)}$$ **Matlab code**:
	 `xISDM=(Rs*Rp+RD*(Rs+Rp))/(Rs+Rp)*I01/(a*Vth)*exp((1/(a*Vth))*(V*(1+RD/Rp)-RD*Ipv+(Rs*Rp+RD*(Rs+Rp))/(Rs+Rp)*(Ipv+I01-V/Rp)))`
	 `I=Rp/(Rs+Rp)*(Ipv+I01-V/Rp-lambertw(xISDM))`

(continued)

Table 2.1 (continued)

Equivalent circuit	Equation and MATLAB code
	$I = I_{PV} + I_0 - \dfrac{V}{R_P} - \dfrac{aV_{th}}{R_S} W\left(x_{SDMR_S}\right)$ $x_{SDMR_S} = \dfrac{R_S I_0}{aV_{th}} e^{\frac{V + R_S(I_{PV} + I_0)}{aV_{th}}}$ **Matlab code**: xSDMRS=Rs*I0/(a*Vth)*exp((V+Rs*(Ipv+I0))/(a*Vth)) I=Ipv+I0-V/Rp-a*Vth/Rs*lambertw(xSDMRS)
	$I = I_{PV} + I_0 - \dfrac{V}{R_P} - \dfrac{aV_{th}}{R_D} W\left(x_{SDMR_D R_P}\right)$ $x_{SDMR_D R_P} = \dfrac{R_D I_0}{aV_{th}} e^{\frac{V + R_D I_0}{aV_{th}}}$ **Matlab code**: xSDMRDRP =RD*I0/(a*Vth)*exp((V+RD*I0)*(a*Vth)) I=Ipv+I0-V/Rp-a*Vth/RD*lambertw(xSDMRDRP)

(continued)

Table 2.1 (continued)

Equivalent circuit	Equation and MATLAB code
	$I = I_{PV} + I_0 - \dfrac{aV_{th}}{R_D+R_S}\, W\left(x_{SDMR_DR_S}\right)$ $$x_{SDMR_DR_S} = \frac{(R_S+R_D)I_0}{aV_{th}}\, e^{\frac{V-R_DI_{PV}+(R_S+R_D)(I_0+I_{PV})}{aV_{th}}}$$ **Matlab code:** `xSDMRDRS =(Rs+RD)*I0/(a*Vth)*exp((V-RD*Ipv+(Rs+Rp)*(Ipv+I0))/(a*Vth))` `I=Ipv+I0-a*Vth/(Rs+RD)*lambertw(xSDMRDRS)`
	$I = I_{SDM} - I_{02}\left(e^{\frac{V}{a_2V_{th}}} - 1\right)$ $$I_{SDM} = \frac{R_P(I_{PV}+I_{01})-V}{R_S+R_P} - \frac{a_1V_{th}}{R_S}\, W\left(\frac{I_{01}R_SR_P}{a_1V_{th}(R_S+R_P)}\, e^{\frac{R_P(R_SI_{PV}+R_SI_{01}+V)}{a_1V_{th}(R_P+R_S)}}\right)$$ **Matlab code:** `x=I01*Rs*Rp/(a1*Vth*(Rs+Rp))*exp((Rp*(Rs*Ipv+Rs*I01+V))/(a1*Vth*(Rs+Rp)));` `ISDM=(Rp*(Ipv+I01)-V)/(Rs+Rp)-a1*Vth*lambertw(x)/Rs;` `I=ISDM-I02*(exp(V/(a2*Vth))-1)`

(continued)

Table 2.1 (continued)

Equivalent circuit	Equation and MATLAB code

(circuit diagram: current source I_{SDM}, diode D, resistor R_{D}, terminals V and I)	$I = I_{\mathrm{SDM}} - I_{\mathrm{D2}}$
	
	 $I_{\mathrm{D2}} = \dfrac{a_2 V_{\mathrm{th}}}{R_{\mathrm{D}}} W\left(\dfrac{R_{\mathrm{D}} I_{02}}{a_2 V_{\mathrm{th}}} e^{\frac{V + R_{\mathrm{D}} I_{02}}{a_2 V_{\mathrm{th}}}} \right) - I_{02}$
	
	 $I_{\mathrm{SDM}} = \dfrac{R_{\mathrm{P}}(I_{\mathrm{PV}} + I_{01}) - V}{R_{\mathrm{S}} + R_{\mathrm{P}}} - \dfrac{a_1 V_{\mathrm{th}}}{R_{\mathrm{S}}} W\left(\dfrac{I_{01} R_{\mathrm{S}} R_{\mathrm{P}}}{a_1 V_{\mathrm{th}}(R_{\mathrm{S}} + R_{\mathrm{P}})} e^{\frac{R_{\mathrm{P}}(R_{\mathrm{S}} I_{\mathrm{PV}} + R_{\mathrm{S}} I_{01} + V)}{a_1 V_{\mathrm{th}}(R_{\mathrm{P}} + R_{\mathrm{S}})}} \right)$
	
	 Matlab code:
	
	 `x=I01*Rs*Rp/(a1*Vth*(Rs+Rp))*exp((Rp*(Rs*Ipv+Rs*I01+V))/(a1*Vth*(Rs+Rp)));`
	 `ISDM=(Rp*(Ipv+I01)-V)/(Rs+Rp)-a1*Vth*lambertw(x)/Rs;`
	 `ID2=a2*Vth/RD*lambertw(RD*I02/(a2*Vth)*exp((V+RD*I02)/(a2*Vth)))-I02`
	 `I=ISDM-ID2`

Compared to the SDM, which involves five basic parameters, the DDM is more complex, as it includes a total of seven independent parameters. In addition to the photo-generated current I_{PV}, the series resistance R_S, and the shunt resistance R_P, the DDM includes two diodes, each with its saturation current (I_{01} and I_{02}) and ideality factor (a_1 and a_2).

This added flexibility enables more accurate modeling of real solar cell behavior, but it also requires advanced parameter identification techniques and careful processing of measurement data to avoid parameter correlation and ensure numerical stability of the solution.

To that end, the model and identification procedure must be carefully designed to ensure:

- each parameter has a clearly defined and independent role,
- the obtained solutions are physically realistic,
- the algorithm operates reliably, without oscillations or incorrect conclusions.

The standard configuration of the DDM includes the following five fundamental components:

- a photo-generated current source I_{PV}—models the current generated by solar radiation absorbed in the active layer of the solar cell;
- the first ideal diode with ideality factor a_1—represents the current governed by recombination processes in the bulk region of the semiconductor junction;
- the second ideal diode with ideality factor a_2—used to model recombination in the depletion region, which is particularly significant at low voltages and in high-performance cells;
- a series resistance R_S—describes ohmic losses within the cell structure, including metallization, contacts, and conductive layers;
- a shunt resistance R_P—simulates unwanted leakage paths and structural imperfections that compromise the insulation of the PN junction.

Figure 2.2 illustrates the electrical schematic of the Double Diode Model (DDM), which includes the photo-generated current source, two ideal diodes, as well as the series and shunt resistances.

As previously mentioned, the DDM represents an extension of the basic Single Diode Model (SDM), with the goal of more accurately describing the physical mechanisms within a solar cell—particularly under conditions where recombination processes are more pronounced. The electrical output of the cell is still described by its current-voltage (I–V) characteristic, but now with an additional diode element that enables a better representation of behavior across the full operating range, including low-irradiance and high-efficiency regions.

Within the DDM framework, the mathematical relationship between output current and voltage remains nonlinear, but becomes more complex due to the inclusion of the second diode with its own parameters (saturation current and ideality factor). This formulation captures a more detailed picture of carrier transport and

Fig. 2.2 DDM

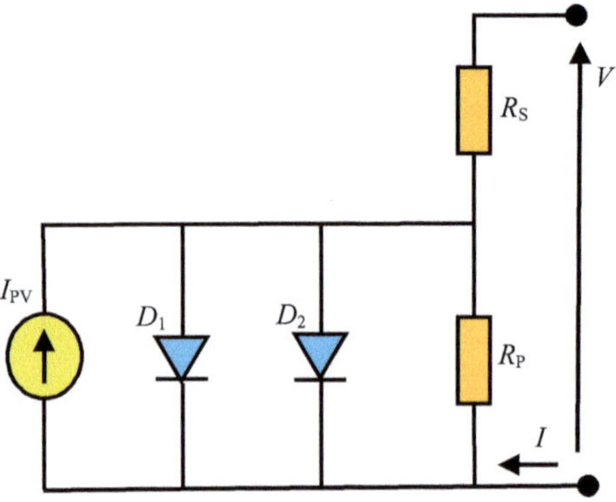

recombination processes in both the bulk region and the depletion region of the PN junction, as well as the internal losses and structural imperfections present within the solar cell.

Following these physical assumptions, the current-voltage characteristic of a solar cell under the DDM model is formally expressed by the following nonlinear equation:

$$I = I_{PV} - I_{01}\left(e^{\frac{V+IR_S}{a_1 V_{th}}} - 1\right)$$
$$- I_{02}\left(e^{\frac{V+IR_S}{a_2 V_{th}}} - 1\right) - \frac{V+IR_S}{R_P}, \tag{2.5}$$

where

I—output current of the solar cell [A],
V—output voltage [V],
I_{PV}—photo-generated current, proportional to the light intensity [A],
I_{01} and I_{02}—diode saturation currents [A],
a_1 and a_2—diode ideality factors (typically between 1 and 2),
R_S—series resistance [Ω],
R_P—shunt (parallel) resistance [Ω],
V_{th}—thermal voltage ($=kBT/q$) [V],
q—elementary charge ($\approx 1.602 \times 10^{-19}$ C),
kB—Boltzmann constant ($\approx 1.381 \times 10^{-23}$ J/K),
T—absolute temperature [K].

Unlike the SDM model, in which a closed-form analytical expression can be derived under certain conditions (e.g., using the Lambert W function), the standard DDM model does not allow such a formulation. The main reason lies in the presence of two exponential functions, both of which have arguments involving both current (I) and voltage (V), significantly complicating the mathematical structure of the expression.

Due to this nonlinear and non-reducible form, the DDM model equation cannot be solved analytically concerning any variable without introducing additional approximations. Therefore, in practice, solving this equation relies on numerical methods or optimization algorithms that iteratively adjust model parameters to achieve agreement with experimental I–V data.

In the available literature [3], an original iterative method based on the application of the Lambert W function can be found. Specifically, the general current–voltage equation for the two-diode configuration can, through suitable mathematical transformations, be reduced to a form amenable to solving using this function:

$$\alpha + \beta e^{\delta y} = y e^{y}, \tag{2.6}$$

where

$$\alpha = \frac{R_S R_P}{R_S + R_P} \frac{I_{01}}{a_1 V_{th}} e^{\frac{V}{a_1 V_{th}}} e^{\frac{R_S R_P}{R_S + R_P} \frac{1}{a_1 V_{th}}\left(I_{PV} + I_{01} + I_{02} - \frac{V}{R_P}\right)}, \tag{2.7}$$

$$\beta = \frac{R_S R_P}{R_S + R_P} \frac{I_{02}}{a_1 V_{th}} e^{\frac{V}{a_2 V_{th}}} e^{\frac{R_S R_P}{R_S + R_P} \frac{1}{a_2 V_{th}}\left(I_{PV} + I_{01} + I_{02} - \frac{V}{R_P}\right)}, \tag{2.8}$$

and

$$\delta = 1 - \frac{a_1}{a_2}. \tag{2.9}$$

The previous equation can be solved via the Lambert W iterative procedure. The MATLAB code for its implementation is given below:

```
for t=1:length(V) % V – voltage vector

ALFA=((Rs/Vth/a1)/(1+Rs/Rp))*(I01*exp(V(t)/Vth/a1))*exp((Rs/Vth/
a1)*(Ipv+I01+I02-V(t)/Rp)/(1+Rs/Rp));

BETA=((Rs/Vth/a1)/(1+Rs/Rp))*(I02*exp(V(t)/Vth/a2))*exp((Rs/Vth/
a2)*(Ipv+I01+I02-V(t)/Rp)/(1+Rs/Rp));

GAMA1-a1/a2;

initial=0;

for counter=1:1000
        current_iteration= counter;
        z=ALFA+BETA*exp(GAMA*initial);
        y=lambertw(z);
    if abs(z-(ALFA+BETA*exp(GAMA*y)))<1e-10
        break
    end
     initial=y;
```

end

x=y/((Rs/Vth/a1)/(1+Rs/Rp));

I(t)=((Ipv+I01+I02-V(t)/Rp)-x)/(1+Rs/Rp);

end

However, the literature offers several approaches for solving the current–voltage equation [14, 15, 20]:

- An approximate method based on the linearization of the exponential terms in the current-voltage equation, aimed at simplifying the solution process. This approach yields an approximate analytical expression, which significantly facilitates its application in the analysis and simulation of photovoltaic systems. Despite being an approximation, the obtained results show a high degree of agreement with measured data.
- Models based on repositioning the series resistor in three different configurations, while the parallel resistor retains its standard location. In this way, each of the proposed topologies enables the derivation of a closed-form analytical solution for the current-voltage characteristic, which considerably simplifies the mathematical treatment and accelerates the simulation procedures.
- A model based on placing the series resistance in a common branch shared by both diodes D_1 and D_2, while the position of the parallel resistor remains unchanged. This significantly improves efficiency as well as the simplicity of deriving the I–V characteristic. However, although this scheme offers advantages in terms of stability and simplification of the I–V function itself, it does not allow for a closed-form analytical solution, and the calculation of the characteristic must rely on iterative numerical methods.

All previously mentioned models are illustrated in Table 2.2, along with the current-voltage equation and the corresponding MATLAB code used to define current as a function of voltage.

For the two-diode model (DDM), it is also important to have an explicit formula for voltage as a function of current, especially when comparing with experimental I–V data or during parameter optimization. However, in this case as well, the application of the Lambert W function encounters the same numerical limitations at low current levels or high values of the parallel resistance, since the arguments of the exponential terms may reach values outside the supported domain of most software packages.

To overcome this issue, in the available literature [21], an approximate analytical invertible expression of the voltage-current (V–I) characteristics for both DDM and TDM models, based on the g-function, can be found.

It has the following form:

$$V = a_1 V_{\text{th}} g(\alpha) - R_S I - a_1 V_{\text{th}} \log\left(\frac{R_P(I_{01} + p_2 I_{02})}{a_1 V_{\text{th}}}\right), \qquad (2.10)$$

where

Table 2.2 DDM—equivalent circuit and solutions

Equivalent circuit	Equation and MATLAB code

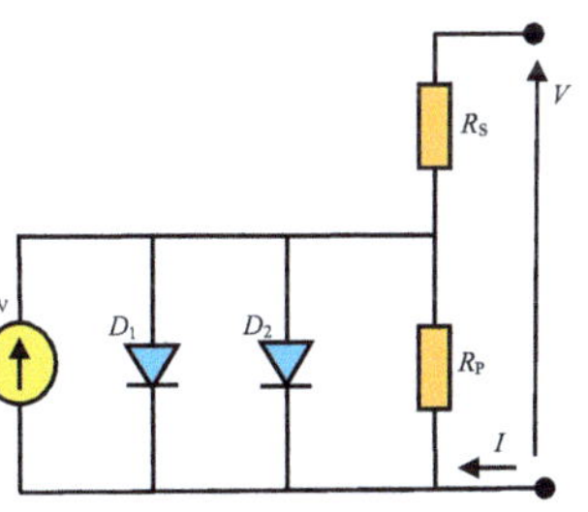	Approximate solution:
	
	$$I = \frac{R_P(I_{PV} + I_{01} + I_{02}) - V}{R_P + R_S}$$
	$$- \frac{a_1 V_{th}}{2R_S} W\left(\frac{R_S R_P (I_{01} + I_{02})}{a_1 V_{th}(R_P + R_S)} e^{\frac{R_P(R_S I_{PV} + R_S I_{01} + R_S I_{02} + V)}{a_1 V_{th}(R_P + R_S)}} \right)$$
	$$- \frac{a_2 V_{th}}{2R_S} W\left(\frac{R_S R_P (I_{01} + I_{02})}{a_2 V_{th}(R_P + R_S)} e^{\frac{R_P(R_S I_{PV} + R_S I_{01} + R_S I_{02} + V)}{a_2 V_{th}(R_P + R_S)}} \right)$$
	
	Matlab code:
	`x1=Rs*Rp*(I01+I02)/`
	`(a1*Vth*(Rs+Rp))*exp((Rp*(Rs*Ipv+Rs*I01+Rs*I02+V(t))))/`
	`(a1*Vth*(Rs+Rp))) x2=Rs*Rp*(I01+I02)/`
	`(a2*Vth*(Rs+Rp))*exp((Rp*(Rs*Ipv+Rs*I01+Rs*I02+V(t))))/`
	`(a2*Vth*(Rs+Rp)))`
	`I=(Rp*(Ipv+I01+I02)-V(t))/(Rs+Rp)-a1*Vth/`
	`(2*Rs)*lambertw(x1)-a2*Vth/(2*Rs)*lambertw(x2);`

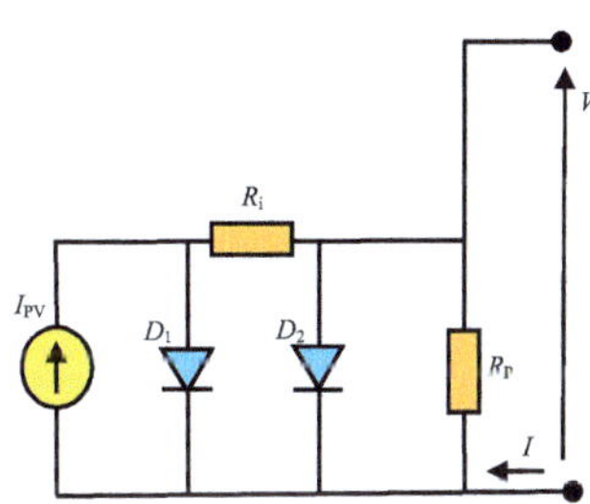	$$I = I_{PV} + I_{01} - I_{02}\left(e^{\frac{V}{a_2 V_{th}}} - 1 \right) - \frac{V}{R_P} - \frac{a_1 V_{th}}{R_i} W\left(x_{DDMRi} \right)$$
	
	$$x_{DDMRi} = \frac{R_i I_{01}}{a_1 V_{th}} e^{\frac{V + R_i(I_{PV} + I_{01})}{a_1 V_{th}}}$$
	Matlab code:
	`xDDMRi =Ri*I01/(a1*Vth)*exp((V+Ri*(Ipv+I01))/`
	`(a1*Vth))`
	`I=Ipv+I01-I02*(exp(V/(a2*Vth))-1) V/Rp-a1*Vth/`
	`Ri*lambertw(xDDMRi)`

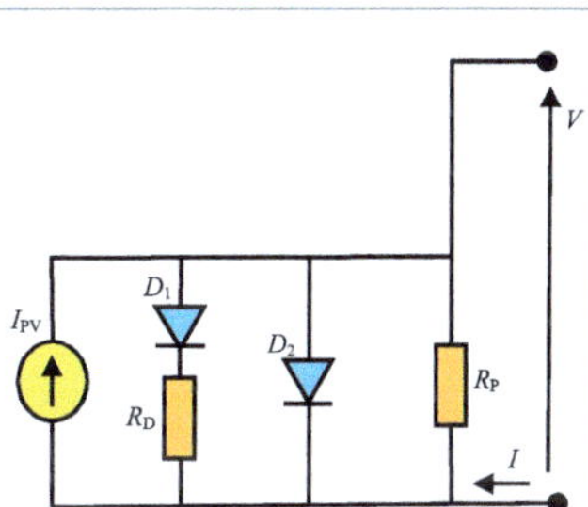	$$I = I_{PV} + I_{01} - I_{02}\left(e^{\frac{V}{a_2 V_{th}}} - 1 \right) - \frac{V}{R_P} - \frac{a_1 V_{th}}{R_D} W\left(x_{DDMR_D} \right)$$
	
	$$x_{DDMR_D} = \frac{R_D I_{01}}{a_1 V_{th}} e^{\frac{V + R_D I_{01}}{a_1 V_{th}}}$$
	Matlab code:
	`xDDMRD=RD*I01/(a1*Vth)*exp((V+RD*I01)/(a1*Vth))`
	`I=Ipv+I01-I02*(exp(V/(a2*Vth))-1)-V/Rp-a1*Vth/`
	`RD*lambert(xDDMRD)`

(continued)

Table 2.2 (continued)

Equivalent circuit	Equation and MATLAB code
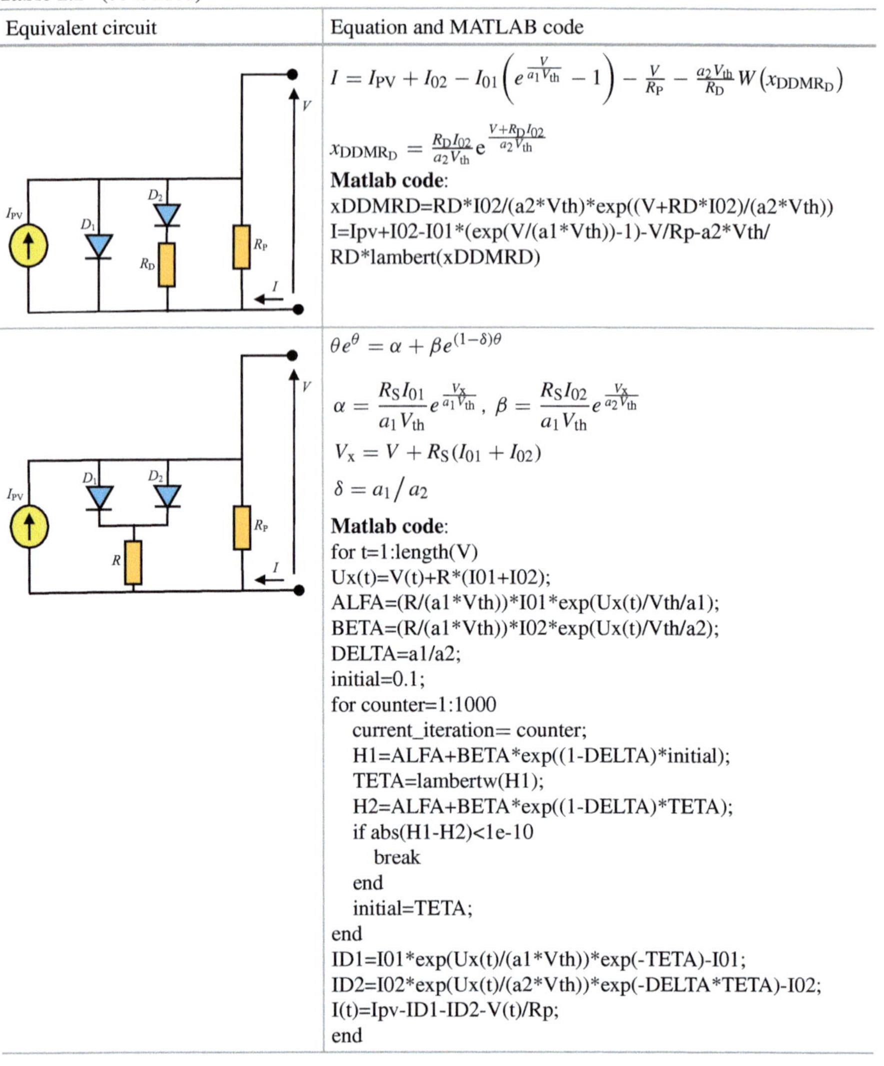	$$I = I_\mathrm{PV} + I_{02} - I_{01}\left(e^{\frac{V}{a_1 V_\mathrm{th}}} - 1\right) - \frac{V}{R_\mathrm{P}} - \frac{a_2 V_\mathrm{th}}{R_\mathrm{D}} W\left(x_{\mathrm{DDMR_D}}\right)$$ $$x_{\mathrm{DDMR_D}} = \frac{R_\mathrm{D} I_{02}}{a_2 V_\mathrm{th}} e^{\frac{V + R_\mathrm{D} I_{02}}{a_2 V_\mathrm{th}}}$$ **Matlab code**: xDDMRD=RD*I02/(a2*Vth)*exp((V+RD*I02)/(a2*Vth)) I=Ipv+I02-I01*(exp(V/(a1*Vth))-1)-V/Rp-a2*Vth/ RD*lambert(xDDMRD)
	$$\theta e^{\theta} = \alpha + \beta e^{(1-\delta)\theta}$$ $$\alpha = \frac{R_\mathrm{S} I_{01}}{a_1 V_\mathrm{th}} e^{\frac{V_\mathrm{x}}{a_1 V_\mathrm{th}}},\ \beta = \frac{R_\mathrm{S} I_{02}}{a_1 V_\mathrm{th}} e^{\frac{V_\mathrm{x}}{a_2 V_\mathrm{th}}}$$ $$V_\mathrm{x} = V + R_\mathrm{S}(I_{01} + I_{02})$$ $$\delta = a_1 / a_2$$ **Matlab code**:

```
for t=1:length(V)
Ux(t)=V(t)+R*(I01+I02);
ALFA=(R/(a1*Vth))*I01*exp(Ux(t)/Vth/a1);
BETA=(R/(a1*Vth))*I02*exp(Ux(t)/Vth/a2);
DELTA=a1/a2;
initial=0.1;
for counter=1:1000
   current_iteration= counter;
   H1=ALFA+BETA*exp((1-DELTA)*initial);
   TETA=lambertw(H1);
   H2=ALFA+BETA*exp((1-DELTA)*TETA);
   if abs(H1-H2)<1e-10
     break
   end
   initial=TETA;
end
ID1=I01*exp(Ux(t)/(a1*Vth))*exp(-TETA)-I01;
ID2=I02*exp(Ux(t)/(a2*Vth))*exp(-DELTA*TETA)-I02;
I(t)=Ipv-ID1-ID2-V(t)/Rp;
end
```

$$\alpha = \frac{R_\mathrm{P}(I_\mathrm{PV} + I_{01} + I_{02} - I)}{a_1 V_\mathrm{th}} + \log\left(\frac{R_\mathrm{P}(I_{01} + p_2 I_{02})}{a_1 V_\mathrm{th}}\right). \tag{2.11}$$

In the previous equation, variable p is as follows:

$$p_2 = \frac{e^{\frac{V + I R_\mathrm{S}}{a_2 V_\mathrm{th}}}}{e^{\frac{V + I R_\mathrm{S}}{a_1 V_\mathrm{th}}}}, \tag{2.12}$$

i.e. it is defined by considering all pairs of measured voltage and current values.

As in the case of the SDM, the introduction of the g-function (also known as the LogWright function) enables a closed-form approximate analytical expression for the voltage over the entire operating current range. Such an analytical form, although approximate, offers high accuracy compared to numerically obtained solutions. Therefore, the application of the g-function in the DDM model has proven to be a highly efficient approach: it enables direct and fast calculation of voltage for a given current, without the need for iterative solving, while maintaining the reliability and precision of the model across the entire range of operating points.

2.2.3 TDM

The Triple Diode Model (TDM) represents a further extension of the two-diode structure, aimed at incorporating additional recombination mechanisms that are not adequately described by the SDM and DDM approaches [3]. This model includes three ideal diodes, each corresponding to a different type of recombination: the first diode models recombination in the bulk region of the PN junction, the second accounts for recombination within the depletion region, while the third diode is introduced to represent more complex phenomena such as surface recombination and recombination effects under high-injection operating conditions. As a result, the TDM model significantly improves the agreement with experimental I–V data across a wide range of illumination intensities and load conditions, particularly in regions where even the DDM fails to ensure sufficient accuracy (e.g., at high voltages near open-circuit conditions).

Compared to the SDM and DDM models, the triple diode model is characterized by even greater complexity, as it involves a total of nine independent parameters. In addition to the photo-generated current I_{PV}, series resistance R_S, and parallel (shunt) resistance R_P, the TDM includes three diodes, each with its saturation current (I_{01}, I_{02}, I_{03}) and ideality factor (a_1, a_2, a_3). This rich parameter space enables the model to be fitted very precisely to actual solar cell characteristics; however, it also necessitates the use of advanced numerical or approximation techniques (e.g., hybrid optimization) to prevent excessive parameter correlation and ensure computational stability.

Due to its comprehensive coverage of diverse recombination mechanisms, the TDM is most commonly used in applications where maximum precision is critical— for example, in the design of high-efficiency photovoltaic modules or in detailed studies of degradation effects under various operating conditions.

The standard configuration of the TDM model consists of the following six basic elements:

- Photo-generated current source I_{PV}: Models the generation of electric current due to the absorption of solar radiation in the active layer of the solar cell.
- First ideal diode with ideality factor a_1: Represents current resulting from recombination processes in the bulk region of the semiconductor junction.

- Second ideal diode with ideality factor a_2: Models recombination in the depletion region, which becomes significant at lower voltages and in high-efficiency cells.
- Third ideal diode with ideality factor a_3: Represents surface recombination and additional recombination effects occurring under high-injection operating regimes.
- Series resistance R_S: Describes ohmic losses within the cell structure, including metallization, contacts, and conductive layers.
- Parallel (shunt) resistance R_P: Simulates unwanted leakage paths and structural imperfections that degrade the isolation of the PN junction.

Figure 2.3 shows the equivalent circuit of the TDM model, which includes the photo-generated current source, three ideal diodes, as well as series and parallel resistances.

The mathematical formulation of the triple diode model (TDM) is significantly more complex compared to the SDM and DDM approaches. As mentioned earlier, TDM introduces three diodes, each with its own parameters, which leads to additional complexity both in terms of structure and analytical treatment. Each of these diodes contributes an exponential term in the current expression, where each term is based on a different ideality factor and saturation current.

In addition, the large number of parameters increases the risk of correlation among them, further complicating the identification process and potentially affecting the stability and reliability of the obtained solution. Therefore, the practical application of the TDM model often depends on the availability of highly accurate experimental data and the use of advanced optimization techniques.

In accordance with this complex physical and mathematical structure, the current–voltage characteristic of the TDM model is formally expressed by the following equation:

$$I = I_{\text{PV}} - I_{01}\left(e^{\frac{V+IR_S}{a_1 V_{\text{th}}}} - 1\right)$$

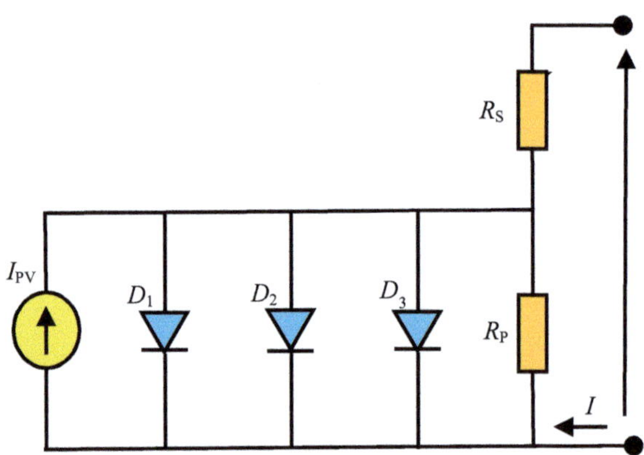

Fig. 2.3 TDM equivalent circuit

$$- I_{02}\left(e^{\frac{V+IR_S}{a_2 V_{th}}} - 1\right) - I_{03}\left(e^{\frac{V+IR_S}{a_3 V_{th}}} - 1\right) - \frac{V + IR_S}{R_P}, \tag{2.13}$$

where

I—output current of the solar cell [A],
V—output voltage [V],
I_{PV}—photo-generated current, proportional to the light intensity [A],
I_{01}, I_{02} and I_{03}—diode saturation currents [A],
a_1, a_2 and a_3—diode ideality factors (typically between 1 and 2),
R_S—series resistance [Ω],
R_P—shunt (parallel) resistance [Ω],
V_{th}—thermal voltage ($=kBT/q$) [V],
q—elementary charge ($\approx 1.602 \times 10^{-19}$ C),
kB—Boltzmann constant ($\approx 1.381 \times 10^{-23}$ J/K),
T—absolute temperature [K].

Due to its inherently complex structure, the triple-diode model does not admit a closed-form analytical solution in any variable—even when employing special functions such as the Lambert W. This limitation arises primarily from the presence of three exponential terms, each with nonlinear dependences on both voltage and current, as well as distinct ideality factors and saturation currents for each diode.

The resulting mathematical structure yields a highly nonlinear and implicit equation, in which the variables are interdependent in a manner that prevents the direct symbolic isolation of any single one. As a result, solving the model requires the use of numerical methods, which allow for iterative determination of the solution based on known input parameters and experimental data.

In the literature [3], an original iterative approach based on the application of the iterative Lambert W function can be found. In this case, after appropriate mathematical manipulations, the equation describing the TDM can be reduced to the following form:

$$\alpha + \beta e^{\delta y} + \gamma e^{\sigma y} = y e^{y}, \tag{2.14}$$

where

$$\alpha = \frac{R_S R_P}{R_S + R_P} \frac{I_{01}}{a_1 V_{th}} e^{\frac{V}{a_1 \cdot V_{th}}} e^{\frac{R_S R_P}{R_S + R_P} \frac{1}{a_1 V_{th}}\left(I_{PV} + I_{01} + I_{02} + I_{03} - \frac{V}{R_P}\right)}, \tag{2.15}$$

$$\beta = \frac{R_S R_P}{R_S + R_P} \frac{I_{02}}{a_1 V_{th}} e^{\frac{V}{a_2 V_{th}}} e^{\frac{R_S R_P}{R_S + R_P} \frac{1}{a_2 V_{th}}\left(I_{PV} + I_{01} + I_{02} + I_{03} - \frac{V}{R_P}\right)}, \tag{2.16}$$

$$\gamma = \frac{R_S R_P}{R_S + R_P} \frac{I_{03}}{a_1 V_{th}} e^{\frac{V}{a_3 V_{th}}} e^{\frac{R_S R_P}{R_S + R_P} \frac{1}{a_3 V_{th}}\left(I_{PV} + I_{01} + I_{02} + I_{03} - \frac{V}{R_P}\right)}, \tag{2.17}$$

$$\delta = 1 - \frac{u_1}{a_2}, \tag{2.18}$$

$$\sigma = 1 - \frac{a_1}{a_3}. \tag{2.19}$$

The MATLAB code for its implementation is given below:

```
for t=1:length(V) % V – voltage vector

ALFA=((Rs/Vth/a1)/(1+Rs/Rp))*(I01*exp(V(t)/Vth/a1))*exp((Rs/Vth/
a1)*(Ipv+I01+I02+I03-V(t)/Rp)/(1+Rs/Rp));

BETA=((Rs/Vth/a1)/(1+Rs/Rp))*(I02*exp(V(t)/Vth/a2))*exp((Rs/Vth/
a2)*(Ipv+I01+I02+I03-V(t)/Rp)/(1+Rs/Rp));

GAMA=((Rs/Vth/a1)/(1+Rs/Rp))*(I03*exp(V(t)/Vth/a3))*exp((Rs/Vth/
a3)*(Ipv+I01+I02+I03-V(t)/Rp)/(1+Rs/Rp));

DELTA=1-a1/a2;

SIGMA=1-a1/a3;

initial=0;

for counter=1:1000
        current_iteration= counter;
        z=ALFA+BETA*exp(DELTA*initial)+ GAMA*exp(SIGMA*initial);
        y=lambertw(z);
    if abs(z-(ALFA+BETA*exp(DELTA*y) + GAMA*exp(SIGMA*y)))<1e-10
      break
    end
    initial=y;
  end

x=y/((Rs/Vth/a1)/(1+Rs/Rp));

I(t)=((Ipv+I01+I02+I03-V(t)/Rp)-x)/(1+Rs/Rp);

end
```

In order to overcome the pronounced nonlinearity and numerical challenges associated with solving the triple diode model equation, the literature presents two approximate solutions. These methods enable efficient current computation without relying on conventional iterative algorithms. Instead of simplifying the model itself, the proposed strategies employ suitable mathematical transformations that lead to explicit expressions, making them well-suited for both analysis and simulation of photovoltaic systems [21].

Moreover, the literature also introduces enhanced configurations of the triple diode model that differ from the conventional formulation in the structural placement of the series resistance R_S [12]. The motivation behind these modifications was not merely structural, but driven by the need to reformulate the current–voltage relationship into a form more amenable to analytical solutions. As a result of the proposed topological adjustments, closed-form expressions for the I–V characteristics were derived, providing a convenient basis for direct implementation in simulation and optimization algorithms. Validation confirmed that the proposed models achieve a high level of agreement with both numerical and experimental results, while also offering the advantage of reduced computational complexity.

All previously mentioned approximated solutions and models are illustrated in Table 2.3, along with the current-voltage equation and the corresponding MATLAB code used to define current as a function of voltage.

For the triple diode model (TDM), as in the previous cases, having an analytical expression that relates voltage to current is of particular importance, especially in the context of parameter identification and model validation based on experimental data. Given the added complexity of this model, the mathematical challenges become even more pronounced, particularly in low-current regimes and at high values of parallel resistance, where conventional Lambert W function-based methods tend to lose numerical stability.

To overcome the aforementioned limitations, in the available literature [10], an approximate analytical expression for voltage as a function of current in the TDM model, derived using the g-function, can be found. This approach enables the formulation of closed-form expressions that remain valid across the entire operating current range, eliminating the need for numerical inversion or iterative procedures. The previous can be expressed as follows:

$$V = a_1 V_{\text{th}} g(\alpha) - R_S I - a_1 V_{\text{th}} \log\left(\frac{R_P(I_{01} + p_2 I_{02} + p_3 I_{03})}{a_1 V_{\text{th}}}\right), \tag{2.20}$$

where

$$\alpha = \frac{R_P(I_{PV} + I_{01} + I_{02} + I_{03} - I)}{a_1 V_{th}} + \log\left(\frac{R_P(I_{01} + p_2 I_{02} + p_3 I_{03})}{a_1 V_{th}}\right). \tag{2.21}$$

In the preceding equations, the variables p_2 and p_3 are defined as follows:

$$
\begin{aligned}
p_2 &= \frac{e^{\frac{V+IR_S}{a_2 V_{\text{th}}}}}{e^{\frac{V+IR_S}{a_1 V_{\text{th}}}}} \\[2ex]
p_3 &= \frac{e^{\frac{V+IR_S}{a_3 V_{\text{th}}}}}{e^{\frac{V+IR_S}{a_1 V_{\text{th}}}}}.
\end{aligned}
\tag{2.22}
$$

Therefore, the variables p_2 and p_3 are defined based on the measured current–voltage data pairs.

This formulation has proven particularly useful in the application of optimization techniques that rely on direct evaluation of voltage values, as well as in simulation environments involving a high number of iterations.

Table 2.3 TDM—equivalent circuit and solutions

Equivalent circuit	Equation and MATLAB code

	Approximate solution—first: $$I = \frac{R_P(I_{PV} + I_{01} + I_{02} + I_{03}) - V}{R_P + R_S}$$ $$- \frac{a_1}{3R_S} W\left(\frac{R_S R_P(I_{01} + I_{02} + I_{03})}{a_1(R_P + R_S)} e^{\frac{R_P[R_S(I_{PV}+I_{01}+I_{02}+I_{03})+V]}{a_1(R_P+R_S)}} \right)$$ $$- \frac{a_2}{3R_S} W\left(\frac{R_S R_P(I_{01} + I_{02} + I_{03})}{a_2(R_P + R_S)} e^{\frac{R_P[R_S(I_{PV}+I_{01}+I_{02}+I_{03})+V]}{a_2(R_P+R_S)}} \right)$$ $$- \frac{a_3}{3R_S} W\left(\frac{R_S R_P(I_{01} + I_{02} + I_{03})}{a_3(R_P + R_S)} e^{\frac{R_P[R_S(I_{PV}+I_{01}+I_{02}+I_{03})+V]}{a_3(R_P+R_S)}} \right)$$ **Matlab code**:
	```x1=Rs*Rp*(I01+I02+I03)/(a1*Vth*(Rs+Rp))*```
	```exp((Rp*(Rs*Ipv+Rs*I01+Rs*I02+Rs*I03+V))/(a1*Vth*(Rs+Rp)));```
	```x2=Rs*Rp*(I01+I02+I03)/(a2*Vth*(Rs+Rp))*```
	```exp((Rp*(Rs*Ipv+Rs*I01+Rs*I02+Rs*I03+V))/(a2*Vth*(Rs+Rp)));```
	```x3=Rs*Rp*(I01+I02+I03)/(a3*Vth*(Rs+Rp))*```
	```exp((Rp*(Rs*Ipv+Rs*I01+Rs*I02+Rs*I03+V))/(a3*Vth*(Rs+Rp)));```
	```I=(Rp*(Ipv+I01+I02+I03)-V)/(Rs+Rp)-```
	```a1*Vth/(3*Rs)*lambertw(x1)-a2*Vth/(3*Rs)*lambertw(x2)-a3*Vth/(3*Rs)*lambertw(x3);```

(continued)

Table 2.3 (continued)

Equivalent circuit	Equation and MATLAB code
I_{PV}, L_1, D_2, D_3, R_S, R_P, V, I	Approximate solution—second: $$I = \frac{R_P(I_{PV} + I_{01} + I_{02} + I_{03}) - V}{R_P + R_S}$$ $$-\frac{a_1}{3R_S}W\left(\frac{R_S R_P(a_1 I_{01} + a_2 I_{02} + a_3 I_{03})}{a_1^2(R_P + R_S)}e^{\frac{R_P[R_S(I_{PV}+R_S I_{01}+R_S I_{02}+R_S I_{03})+V]}{a_1(R_P+R_S)}}\right)$$ $$-\frac{a_2}{3R_S}W\left(\frac{R_S R_P(a_1 I_{01} + a_2 I_{02} + a_3 I_{03})}{a_2^2(R_P + R_S)}e^{\frac{R_P[R_S(I_{PV}+R_S I_{01}+R_S I_{02}+R_S I_{03})+V]}{a_2(R_P+R_S)}}\right)$$ $$-\frac{a_3}{3R_S}W\left(\frac{R_S R_P(a_1 I_{01} + a_2 I_{02} + a_3 I_{03})}{a_3^2(R_P + R_S)}e^{\frac{R_P[R_S(I_{PV}+R_S I_{01}+R_S I_{02}+R_S I_{03})+V]}{a_3(R_P+R_S)}}\right)$$ **Matlab code:** ```x1=Rs*Rp*(a1*Vth*I01+a2*Vth*I02+a3*Vth*I03)/(((a1*Vth)^2)*(Rs+Rp))*exp((Rp*(Rs*Ipv+Rs*I01+Rs*I02+Rs*I03+V))/(a1*Vth*(Rs+Rp)));x2=Rs*Rp*(a1*Vth*I01+a2*Vth*I02+a3*Vth*I03)/(((a2*Vth)^2)*(Rs+Rp))*exp((Rp*(Rs*Ipv+Rs*I01+Rs*I02+Rs*I03+V))/(a2*Vth*(Rs+Rp)));x3=Rs*Rp*(a1*Vth*I01+a2*Vth*I02+a3*Vth*I03)/(((a3*Vth)^2)*(Rs+Rp))*exp((Rp*(Rs*Ipv+Rs*I01+Rs*I02+Rs*I03+V))/(a3*Vth*(Rs+Rp)));I=(Rp*(Ipv+I01+I02+I03)-V)/(Rs+Rp)-a1*Vth/(3*Rs)*lambertw(x1)-a2*Vth/(3*Rs)*lambertw(x2)-a3*Vth/(3*Rs)*lambertw(x3);```

(continued)

Table 2.3 (continued)

Equivalent circuit	Equation and MATLAB code
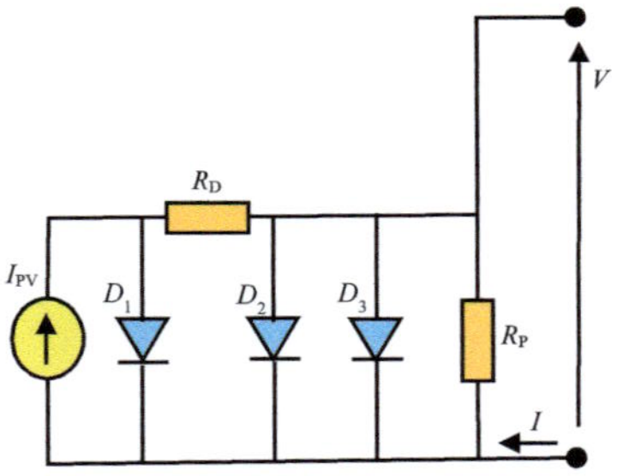 (equivalent circuit diagram)	$I = I_{\mathrm{PV}} + I_{01} - I_{\mathrm{D2}} - I_{\mathrm{D3}} - \dfrac{V}{R_{\mathrm{P}}} - \dfrac{a_1 V_{\mathrm{th}}}{R_{\mathrm{D}}} W(x_{\mathrm{TDM\text{-}M1}})$ $x_{\mathrm{TDM\text{-}M1}} = \dfrac{R_{\mathrm{D}} I_{01}}{a_1 V_{\mathrm{th}}} e^{\frac{V + R_{\mathrm{D}}(I_{\mathrm{PV}} + I_{01})}{a_1 V_{\mathrm{th}}}}$ **Matlab code:** xTDM1=((I01)*RD/(a1*Vth))*exp((V+RD*(Ipv+I01))/(a1*Vth)); I=Ipv+(I01)-(I02)*(exp(V/(a2*Vth))-1)-(I03)* (exp(V/(a3*Vth))-1)-V/Rp-a1*Vth*lambertw(xTDM1)/RD;
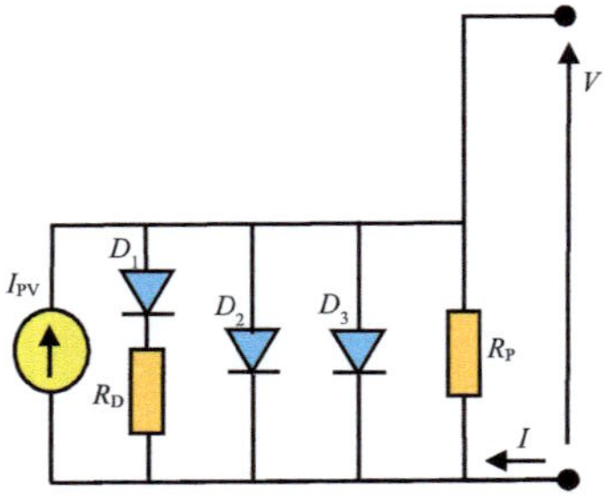 (equivalent circuit diagram)	$I = I_{\mathrm{PV}} + I_{01} - I_{\mathrm{D2}} - I_{\mathrm{D3}} - \dfrac{V}{R_{\mathrm{P}}} - \dfrac{a_1 V_{\mathrm{th}}}{R_{\mathrm{D}}} W(x_{\mathrm{TDM\text{-}M2}})$ $x_{\mathrm{TDM\text{-}M2}} = \dfrac{R_{\mathrm{D}} I_{01}}{a_1 V_{\mathrm{th}}} e^{\frac{V + R_{\mathrm{D}} I_{01}}{a_1 V_{\mathrm{th}}}}$ **Matlab code:** xTDM2=((I01)*RD/(a1*Vth))*exp((V+RD*I01)/(a1*Vth)); I=Ipv+(I01)-(I02)*(exp(V/(a2*Vth))-1)-(I03)* (exp(V/(a3*Vth))-1)-V(t)/Rp-a1*Vth*lambertw(xTDM2)/RD;

2.2.4 FDM

The Four-Diode Model (FDM) represents the most advanced and complex variant of equivalent circuit models for solar cells, developed with the aim of capturing a broader range of physical mechanisms that affect the form and behavior of the current–voltage characteristic [6]. This model is derived by further extending previous concepts through the inclusion of a fourth ideal diode, thereby enabling a high degree of flexibility in modeling complex effects within the PN junction.

In the FDM structure, each of the four diodes has its parameters (ideality factor and saturation current), which enables a precise separation and description of different recombination mechanisms. The first diode is typically associated with recombination in the bulk region, the second with recombination in the depletion region, the third with surface effects and high-injection conditions, while the fourth models recombination through deep energy traps, as well as thermionic emission effects that occur under specific operating conditions.

Compared to the SDM, DDM, and TDM models, the Four-Diode Model exhibits an even higher level of complexity, as it involves a total of eleven independent parameters. In addition to the photo-generated current I_{PV}, the series resistance R_S, and the parallel (shunt) resistance R_P, the FDM includes four diodes, each with its saturation current (I_{01}, I_{02}, I_{03}, and I_{04}) and ideality factor (a_1, a_2, a_3, and a_4).

This configuration enables the FDM to accurately track experimental I–V characteristics even under complex conditions, including extreme levels of irradiance and temperature, as well as degradation processes resulting from aging or mechanical damage to the solar cell. Owing to its high flexibility and rich parameter space, the FDM is frequently employed in advanced performance studies, diagnostics, and reliability testing of high-efficiency photovoltaic modules.

The standard configuration of the FDM model consists of seven fundamental elements:

- Photo-generated current source I_{PV}: Models the generation of electric current due to the absorption of solar radiation in the active layer of the solar cell.
- First ideal diode with ideality factor a_1: Represents current resulting from recombination processes in the bulk region of the semiconductor junction.
- Second ideal diode with ideality factor a_2: Models recombination in the depletion region, which becomes significant at lower voltages and in high-efficiency cells.
- Third ideal diode with ideality factor a_3: Represents surface recombination and additional recombination effects occurring under high-injection operating regimes.
- The fourth ideal diode, characterized by the ideality factor a_4, models recombination through deep energy traps, as well as the effects of thermionic emission phenomena
- Series resistance R_S: Describes ohmic losses within the cell structure, including metallization, contacts, and conductive layers.
- Parallel (shunt) resistance R_P: Simulates unwanted leakage paths and structural imperfections that degrade the isolation of the PN junction.

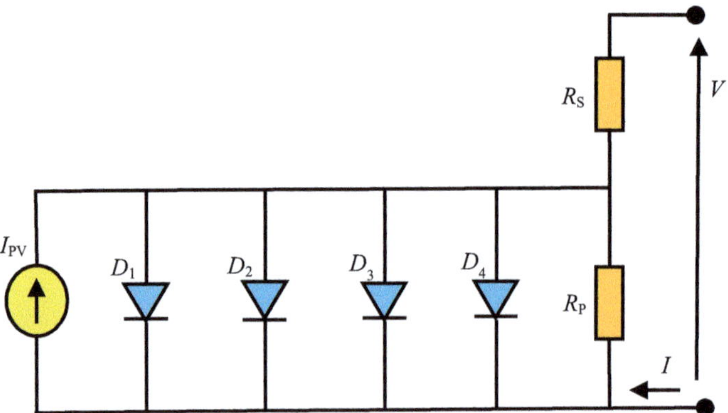

Fig. 2.4 FDM equivalent circuit

Figure 2.4 presents the equivalent circuit diagram of the FDM model, comprising a photo-generated current source, four ideal diodes, and both series and parallel resistances.

Unlike the previous models, the Four-Diode Model (FDM) represents the most complex equivalent structure of a solar cell, both in terms of the physical mechanisms it describes and the mathematical treatment it requires. In the FDM approach, each of the four ideal diodes introduces its own exponential term into the output current expression, with the characteristics of each diode defined by a distinct ideality factor and saturation current.

Such a structure enables a highly detailed differentiation and modeling of various recombination effects, including recombination in the bulk region of the semiconductor and the depletion region, as well as surface and interface effects, and extending to processes associated with high-injection conditions, deep energy traps, and thermionic emissions.

However, this level of accuracy comes at the cost of increased model complexity, as the FDM now comprises eleven parameters. The large number of free parameters significantly raises the risk of mutual correlation, thereby complicating the identification process and requiring a carefully designed optimization environment.

Moreover, due to the presence of four exponential nonlinear terms, analytical solutions of the FDM equation are virtually infeasible without introducing additional assumptions or approximations. Consequently, in practical applications, numerical methods or metaheuristic algorithms are commonly employed to ensure stable and reliable parameter estimation, even in the presence of measurement noise or nonstandard operating conditions.

Formally, the current–voltage characteristic of the FDM model is expressed by the following nonlinear relation:

$$I = I_{\mathrm{PV}} - I_{01}\left(e^{\frac{V+IR_{\mathrm{S}}}{a_1 V_{\mathrm{th}}}} - 1\right) - I_{02}\left(e^{\frac{V+IR_{\mathrm{S}}}{a_2 V_{\mathrm{th}}}} - 1\right)$$

$$-I_{03}\left(e^{\frac{V+IR_S}{a_3 V_{th}}}-1\right)-I_{04}\left(e^{\frac{V+IR_S}{a_4 V_{th}}}-1\right)-\frac{V+IR_S}{R_P},\tag{2.23}$$

where

I—output current of the solar cell [A],
V—output voltage [V],
I_{PV}—photo-generated current, proportional to the light intensity [A],
I_{01}, I_{02}, I_{03} and I_{04}—diode saturation currents [A],
a_1, a_2, a_3, and a_4—diode ideality factors (typically between 1 and 2),
R_S—series resistance [Ω],
R_P—shunt (parallel) resistance [Ω],
V_{th}—thermal voltage ($=kBT/q$) [V],
q—elementary charge ($\approx 1.602 \times 10^{-19}$ C),
kB—Boltzmann constant ($\approx 1.381 \times 10^{-23}$ J/K),
T—absolute temperature [K].

As in the case of the DDM and TDM, the FDM model also lacks an analytical solution. In the available literature, an iterative procedure based on the application of the Lambert W function can be found, which relies on transforming the previously written current–voltage equation into the following form [22]:

$$xe^x = \alpha + \beta e^{\delta x} + \gamma e^{\sigma x} + \varphi e^{\lambda x},\tag{2.24}$$

where

$$\alpha = \frac{R_S R_P}{R_S + R_P}\frac{I_{01}}{a_1 V_{th}}e^{\frac{V}{a_1 V_{th}}}\,e^{\frac{R_S R_P}{R_S+R_P}\frac{1}{a_1 V_{th}}\left(I_{PV}+I_{01}+I_{02}+I_{03}+I_{04}-\frac{V}{R_P}\right)},\tag{2.25}$$

$$\beta = \frac{R_S R_P}{R_S + R_P}\frac{I_{02}}{a_1 V_{th}}e^{\frac{V}{a_2 V_{th}}}\,e^{\frac{R_S R_P}{R_S+R_P}\frac{1}{a_2 V_{th}}\left(I_{PV}+I_{01}+I_{02}+I_{03}+I_{04}-\frac{V}{R_P}\right)},\tag{2.26}$$

$$\gamma = \frac{R_S R_P}{R_S + R_P}\frac{I_{03}}{a_1 V_{th}}e^{\frac{V}{a_3 V_{th}}}\,e^{\frac{R_S R_P}{R_S+R_P}\frac{1}{a_3 V_{th}}\left(I_{PV}+I_{01}+I_{02}+I_{03}+I_{04}-\frac{V}{R_P}\right)},\tag{2.27}$$

$$\varphi = \frac{R_S R_P}{R_S + R_P}\frac{I_{04}}{a_1 V_{th}}e^{\frac{V}{a_4 V_{th}}}\,e^{\frac{R_S R_P}{R_S+R_P}\frac{1}{a_4 V_{th}}\left(I_{PV}+I_{01}+I_{02}+I_{03}+I_{04}-\frac{V}{R_P}\right)},\tag{2.28}$$

$$\delta = 1 - \frac{a_1}{a_2},\tag{2.29}$$

$$\sigma = 1 - \frac{a_1}{a_3},\tag{2.30}$$

$$\lambda = 1 - \frac{a_1}{a_4}\tag{2.31}$$

MATLAB code for solving the mentioned iterative procedure is as follows:

```
for t=1:length(V)

ALFA=((Rs/Vth/a1)/(1+(Rs/Rp)))*((I01*exp(V(t)/Vth/a1))*exp((Rs/Vth/
a1)*(Ipv+I01+I02+I03+I04-V(t)/Rp)/(1+(Rs/Rp))));

BETA=((Rs/Vth/a1)/(1+(Rs/Rp)))*((I02*exp(V(t)/Vth/a2))*exp((Rs/Vth/
a2)*(Ipv+I01+I02+I03+I04-V(t)/Rp)/(1+(Rs/Rp))));

GAMA=((Rs/Vth/a1)/(1+(Rs/Rp)))*((I03*exp(V(t)/Vth/a3))*exp((Rs/Vth/
a3)*(Ipv+I01+I02+I03+I04-V(t)/Rp)/(1+(Rs/Rp))));

FIII=((Rs/Vth/a1)/(1+(Rs/Rp)))*((I04*exp(V(t)/Vth/a4))*exp((Rs/Vth/
a4)*(Ipv+I01+I02+I03+I04-V(t)/Rp)/(1+(Rs/Rp))));

DELTA=1-a1/(a2);

TETA=1-a1/a3;

LAMDA=1-a1/a4;

initial=0;

for count=1:1000
    z=ALFA+BETA*exp(DELTA*initial
)+GAMA*exp(TETA*initial)+FIII*exp(LAMDA*initial);

y=lambertw(z);
    if                      abs(z-(ALFA+BETA*exp(DELTA*y)+GAMA*exp(TETA
*y)+FIII*exp(LAMDA*y)))<1*10^-30
      break

end

initial=y;

end

x=y/((Rs/Vth/a1)/(1+(Rs/Rp)));

I(t)=((Ipv+I01+I02+I03+I04-V(t)/Rp)-x)/(1+(Rs/Rp));

end
```

To accelerate the solution of the current–voltage equations, the following approaches can also be found [5, 6]:

- an approach based on approximate solutions, in which the mathematical expression of the current–voltage characteristic is simplified. These approaches provide a closed-form analytical solution, significantly reducing computational complexity and accelerating the simulation process.
- as well as the application of modified equivalent circuits, aiming to achieve an analytical closed-form solution. As a result, compact models were obtained that allow direct expression of the current–voltage relationship, suitable for rapid implementation in simulation platforms and optimization processes. Analyses

have shown that such configurations achieve a high level of agreement with reference numerical and experimental data, while also contributing to a reduction in the computational complexity of the model.

The mentioned FDMs are presented in the Table 2.4.

Finally, obtaining voltage as a function of current is particularly important for parameter identification, as it allows direct integration into algorithms based on experimental I–V data. Moreover, this approach eliminates the need for numerical inversion and supports fast and stable evaluation, which is crucial for simulation platforms and optimization routines. For the FDM equivalent circuit, the voltage V as a function of output current can be expressed by the following analytical relation:

$$V = a_1 V_{\text{th}} g(\alpha) - R_S I$$
$$- a_1 V_{\text{th}} \log\left(\frac{R_P(I_{01} + p_2 I_{02} + p_3 I_{03} + p_4 I_{04})}{a_1 V_{\text{th}}}\right), \tag{2.32}$$

where

$$\alpha = \frac{R_P(I_{PV} + I_{01} + I_{02} + I_{03} + I_{04} - I)}{a_1 V_{\text{th}}}$$
$$+ \log\left(\frac{R_P(I_{01} + p_2 I_{02} + p_3 I_{03} + p_4 I_{04})}{a_1 V_{\text{th}}}\right). \tag{2.33}$$

In the preceding equations, the variables p_2, p_3, and p_4 are defined as follows (based on measured current–voltage pairs):

$$p_2 = \frac{e^{\frac{V + IR_S}{a_2 V_{\text{th}}}}}{e^{\frac{V + IR_S}{a_1 V_{\text{th}}}}}$$
$$p_3 = \frac{e^{\frac{V + IR_S}{a_3 V_{\text{th}}}}}{e^{\frac{V + IR_S}{a_1 V_{\text{th}}}}} \tag{2.34}$$
$$p_4 = \frac{e^{\frac{V + IR_S}{a_4 V_{\text{th}}}}}{e^{\frac{V + IR_S}{a_1 V_{\text{th}}}}}.$$

2.3 Numerical Examples

This section presents numerical results for the current–voltage (I–V) and voltage–current (V–I) characteristics calculated using the Single Diode Model (SDM) for the RTC France solar cell.

Table 2.4 FDM—equivalent circuit and solutions

Equivalent circuit	Equation and MATLAB code
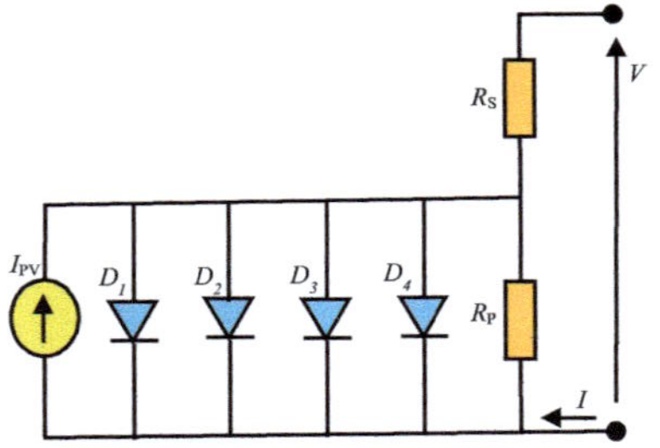	Approximate solutions—first: $$I = \frac{R_P(I_{PV} + I_{01} + I_{02} + I_{03} + I_{04}) - V}{R_P + R_S}$$ $$-\frac{a_1}{4R_S} W\left(\frac{R_S R_P(I_{01} + I_{02} + I_{03} + I_{04})}{a_1(R_P + R_S)} e^{\frac{R_P[R_S(I_{PV}+I_{01}+I_{02}+I_{03}+I_{04})+V]}{a_1(R_P+R_S)}} \right)$$ $$-\frac{a_2}{4R_S} W\left(\frac{R_S R_P(I_{01} + I_{02} + I_{03} + I_{04})}{a_2(R_P + R_S)} e^{\frac{R_P[R_S(I_{PV}+I_{01}+I_{02}+I_{03}+I_{04})+V]}{a_2(R_P+R_S)}} \right)$$ $$-\frac{a_3}{4R_S} W\left(\frac{R_S R_P(I_{01} + I_{02} + I_{03} + I_{04})}{a_3(R_P + R_S)} e^{\frac{R_P[R_S(I_{PV}+I_{01}+I_{02}+I_{03}+I_{04})+V]}{a_3(R_P+R_S)}} \right)$$ $$-\frac{a_4}{4R_S} W\left(\frac{R_S R_P(I_{01} + I_{02} + I_{03} + I_{04})}{a_4(R_P + R_S)} e^{\frac{R_P[R_S(I_{PV}+I_{01}+I_{02}+I_{03}+I_{04})+V]}{a_4(R_P+R_S)}} \right)$$ **Matlab code:** ```x1=Rs*Rp*(I01+I02+I03+I04)/(a1*Vth*(Rs+Rp))* exp((Rp*(Rs*Ipv+Rs*I01+Rs*I02+Rs*I03+Rs*I04+V))/(a1*Vth*(Rs+Rp))); x2=Rs*Rp*(I01+I02+I03+I04)/(a2*Vth*(Rs+Rp))* exp((Rp*(Rs*Ipv+Rs*I01+Rs*I02+Rs*I03+Rs*I04+V))/(a2*Vth*(Rs+Rp))); x3=Rs*Rp*(I01+I02+I03+I04)/(a3*Vth*(Rs+Rp))* exp((Rp*(Rs*Ipv+Rs*I01+Rs*I02+Rs*I03+Rs*I04+V))/(a3*Vth*(Rs+Rp))); x4=Rs*Rp*(I01+I02+I03+I04)/(a4*Vth*(Rs+Rp))* exp((Rp*(Rs*Ipv+Rs*I01+Rs*I02+Rs*I03+Rs*I04+V))/(a4*Vth*(Rs+Rp))); I=(Rp*(Ipv+I01+I02+I03+I04)-V)/(Rs+Rp)-a1*Vth/(4*Rs)*lambertw(x1)-a2*Vth/ (4*Rs)*lambertw(x2)-a3*Vth/(4*Rs)*lambertw(x3)-a4*Vth/(4*Rs)*lambertw(x4);```

(continued)

Table 2.4 (continued)

Equivalent circuit	Equation and MATLAB code
	Approximate solutions—second: $$I = \frac{R_P(I_{PV} + I_{01} + I_{02} + I_{03} + I_{04}) - V}{R_P + R_S}$$ $$- \frac{a_1}{4R_S} W\left(\frac{R_S R_P(a_1 I_{01} + a_2 I_{02} + a_3 I_{03} + a_4 I_{04})}{a_1^2(R_P + R_S)} e^{\frac{R_P[R_S(I_{PV} + R_S I_{01} + R_S I_{02} + R_S I_{03} + a_4 I_{04}) + V]}{a_1(R_P + R_S)}}\right)$$ $$- \frac{a_2}{4R_S} W\left(\frac{R_S R_P(a_1 I_{01} + a_2 I_{02} + a_3 I_{03} + a_4 I_{04})}{a_2^2(R_P + R_S)} e^{\frac{R_P[R_S(I_{PV} + R_S I_{01} + R_S I_{02} + R_S I_{03} + a_4 I_{04}) + V]}{a_2(R_P + R_S)}}\right)$$ $$- \frac{a_3}{4R_S} W\left(\frac{R_S R_P(a_1 I_{01} + a_2 I_{02} + a_3 I_{03} + a_4 I_{04})}{a_3^2(R_P + R_S)} e^{\frac{R_P[R_S(I_{PV} + R_S I_{01} + R_S I_{02} + R_S I_{03} + a_4 I_{04}) + V]}{a_3(R_P + R_S)}}\right)$$ $$- \frac{a_4}{4R_S} W\left(\frac{R_S R_P(a_1 I_{01} + a_2 I_{02} + a_3 I_{03} + a_4 I_{04})}{a_4^2(R_P + R_S)} e^{\frac{R_P[R_S(I_{PV} + R_S I_{01} + R_S I_{02} + R_S I_{03} + a_4 I_{04}) + V]}{a_4(R_P + R_S)}}\right)$$ **Matlab code**: (see below)

Matlab code:

```
x1=Rs*Rp*(a1*Vth*I01+a2*Vth*I02+a3*Vth*I03+a4*Vth*I04)/(((a1*Vth)^2)*(Rs+Rp))*
exp((Rp*(Rs*Ipv+Rs*I01+Rs*I02+Rs*I03+Rs*I04+V))/(a1*Vth*(Rs+Rp)));
x2=Rs*Rp*(a1*Vth*I01+a2*Vth*I02+a3*Vth*I03+a4*Vth*I04)/(((a2*Vth)^2)*(Rs+Rp))*
exp((Rp*(Rs*Ipv+Rs*I01+Rs*I02+Rs*I03+Rs*I04+V))/(a2*Vth*(Rs+Rp)));
x3=Rs*Rp*(a1*Vth*I01+a2*Vth*I02+a3*Vth*I03+a4*Vth*I04)/(((a3*Vth)^2)*(Rs+Rp))*
exp((Rp*(Rs*Ipv+Rs*I01+Rs*I02+Rs*I03+Rs*I04+V))/(a3*Vth*(Rs+Rp)));
x4=Rs*Rp*(a1*Vth*I01+a2*Vth*I02+a3*Vth*I03+a4*Vth*I04)/(((a4*Vth)^2)*(Rs+Rp))*
exp((Rp*(Rs*Ipv+Rs*I01+Rs*I02+Rs*I03+Rs*I04+V))/(a4*Vth*(Rs+Rp)));
I=(Rp*(Ipv+I01+I02+I03+I04)-V)/(Rs+Rp)-a1*Vth/(4*Rs)*lambertw(alfa111)-a2*Vth/
(4*Rs)*lambertw(alfa222)-a3*Vth/(4*Rs)*lambertw(alfa333)-a4*Vth/(4*Rs)*lambertw(alfa444);
```

(continued)

Table 2.4 (continued)

Equivalent circuit	Equation and MATLAB code
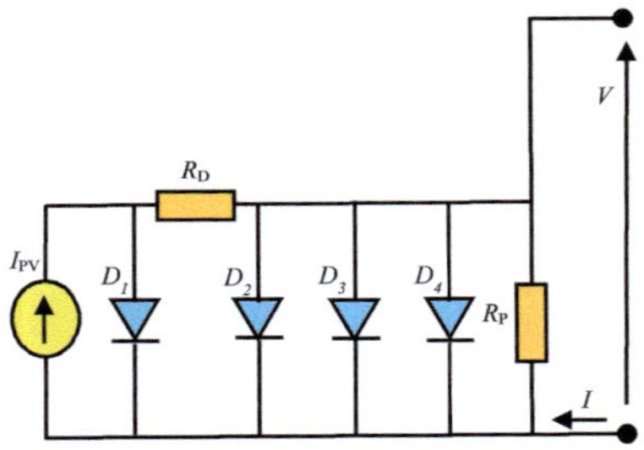	$$I = I_{PV} + I_{01} - I_{02}\left(e^{\frac{V}{a_2 V_{th}}} - 1\right) - I_{03}\left(e^{\frac{V}{a_3 V_{th}}} - 1\right)$$ $$-I_{04}\left(e^{\frac{V}{a_4 V_{th}}} - 1\right) - \frac{V}{R_P} - \frac{a_1 V_{th}}{R_D} W(x_{FDM\text{-}PEC1})$$ $$x_{FDM\text{-}PEC1} = \frac{R_D I_{01}}{a_1 V_{th}} e^{\frac{V + R_D (I_{PV} + I_{01})}{a_1 \cdot V_{th}}}$$ **Matlab code**: ```xFDMPEC1=(I01*RD/(a1*Vth))*exp((V+RD*(Ipv+I01))/(a1*Vth)); I=Ipv+(I01)-(I02)*(exp(V/(a2*Vth))-1)-(I03)*(exp(V/(a3*Vth))-1)-(I04)*(exp(V/(a4*Vth))-1)-V/Rp-a1*Vth*lambertw(xFDMPEC1)/RD;```
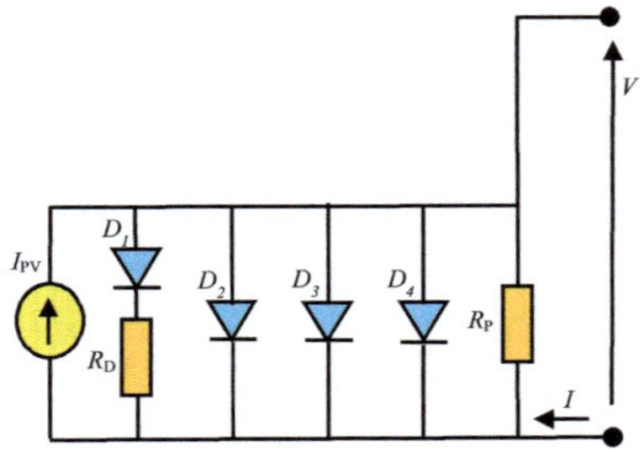	$$I = I_{PV} + I_{01} - I_{02}\left(e^{\frac{V}{a_2 V_{th}}} - 1\right) - I_{03}\left(e^{\frac{V}{a_3 V_{th}}} - 1\right)$$ $$-I_{04}\left(e^{\frac{V}{a_4 V_{th}}} - 1\right) - \frac{V}{R_P} - \frac{a_1 V_{th}}{R_D} W(x_{FDM\text{-}PEC2})$$ $$x_{FDM\text{-}PEC2} = \frac{R_D I_{01}}{a_1 V_{th}} e^{\frac{V + R_D I_{01}}{a_1 V_{th}}}$$ **Matlab code**: ```xFDMPEC2=(I01*RD/(a1*Vth))*exp((V+RD*I01)/(a1*Vth)); I=Ipv+(I01)-(I02)*(exp(V/(a2*Vth))-1)-(I03)*(exp(V/(a3*Vth))-1)-(I04*exp(V/(a4*Vth))-1)-V/Rp-a1*Vth*lambertw(xFDMPEC2)/RD;```

The RTC France solar cell is a well-known solar cell with an open-circuit voltage of 0.6 V, short-circuit current of 0.76 A, current at maximum power of 0.68 A, and voltage at maximum power of 0.48 V, and at temperature of 33^0C.

The measured current–voltage characteristic values are as follows:

Imeasured = [0.7640 0.7620 0.7605 0.7605 0.7600 0.7590 0.7570 0.7570 0.7555 0.7540 0.7505 0.7465 0.7385 0.7280 0.7065 0.6755 0.6320 0.5730 0.4990 0.4130 0.3165 0.2120 0.1035 -0.0100 -0.1230 -0.2100];

Umeasured = [-0.2057 -0.1291 -0.0588 0.0057 0.0646 0.1185 0.1678 0.2132 0.2545 0.2924 0.3269 0.3585 0.3873 0.4137 0.4373 0.4590 0.4784 0.4960 0.5119 0.5265 0.5398 0.5521 0.5633 0.5736 0.5833 0.5900];

To test the application of analytical current–voltage and voltage–current approaches, represented through the Lambert W function and g-function, respectively, the following solar cell parameters were used:

$I_{PV} = 0.760775$;

$I0 = 0.32302$;

$a = 1.4818$;

$R_S = 0.036371$;

$R_P = 53.785$;

Below is the MATLAB code for determining the current–voltage characteristics using both models, along with the corresponding current–voltage and power-current characteristics (Fig. 2.5).

T=33+273.15;

k=1.3806503*10^-23;

q=1.60217646*10^-19;

Vth=k*T/q;

```
% I-V approach
    for t=1:length(Vmeasured)
        x=I0*Rs*Rp/(a*Vth*(Rs+Rp))*exp((Rp*(Rs*Ipv+Rs*I0+Vmeasured(t)))/
(a*Vth*(Rs+Rp)));
        current=(Rp*(Ipv+I0)-Vmeasured(t))/(Rs+Rp)-a*Vth*lambertw(x)/Rs;
        plotcurrent(t)=current;
    end
```

figure(1)

plot(Imeasured, Vmeasured,'kx','Linewidth',2)

hold on

hold on

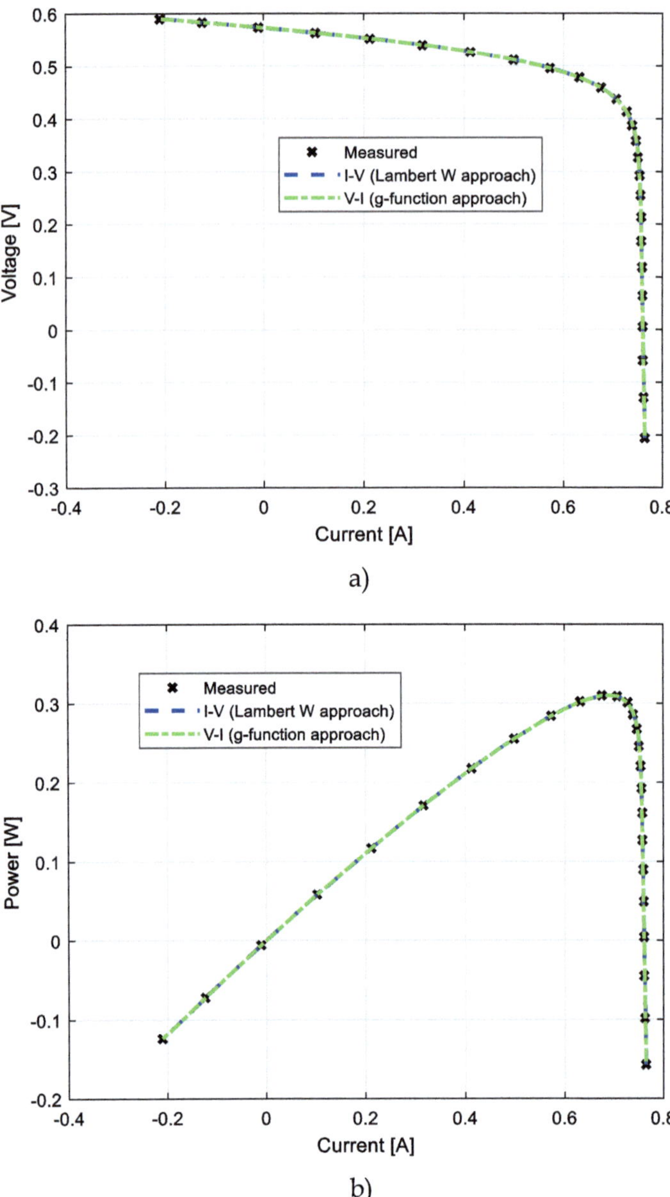

Fig. 2.5 RTC France solar cell, **a** current-voltage and **b** power-current characteristics

```matlab
plot(plotcurrent,Vmeasured ,'b--','Linewidth',2)

hold on

figure (2)

plot(Imeasured, Imeasured.*Vmeasured,'kx','Linewidth',2)

hold on

hold on

plot(plotcurrent, plotcurrent.*Vmeasured,'b--','Linewidth',2)

hold on

% V-I approach
    for t=1:length(Imeasured)    % -10
       alfa=log(I0*Rp/(a*Vth))+((Rp*(Ipv+I0-Imeasured(t)))/(a*Vth));
       if alfa<-exp(1)
           y0=alfa;
         end
         if alfa>exp(1)
           y0=log(alfa);
         end
         if alfa<exp(1) && alfa>=-exp(1)
           y0=-exp(1)+(1+exp(1))/(2*exp(1))*(alfa+exp(1));
         end
         epsilon=10^-15;
         for counter=1:10000
                    y1=y0-(2*(y0+exp(y0)-alfa)*(1+exp(y0)))/(2*(1+exp(y0))^2-
(y0+exp(y0)-alfa)*exp(y0));
             if abs(y1-y0)<epsilon
                break
             end
             y0=y1;
           end
       gfunction=(y1);
       voltage(t)=a*Vth*gfunction-Rs*Imeasured(t)-a*Vth*log(Rp*Io*(10^(-6))/
(a*Vth));
    end
       figure(1)
    hold on
    plot(Imeasured,voltage, 'g-.','Linewidth',2)
    hold on
    grid on
    box on
    xlabel('Current [A]')
    ylabel('Voltage [V]')
```

```
legend('Measured','I-V (Lambert W approach)','V-I (g-function approach)')
 figure(2)
hold on
plot(Imeasured,voltage.*Imeasured, 'g-.','Linewidth',2)
hold on
grid on
box on
xlabel('Current [A]')
ylabel('Power [W]')
legend('Measured','I-V (Lambert W approach)','V-I (g-function approach)')
```

References

1. Younesi A, Shayeghi H, Wang Z, Siano P, Sani A, Safari A (2022) Trends in modern power systems resilience: state-of-the-art review. Renew Sustain Energy Rev 162:112397. https://doi.org/10.1016/j.rser.2022.112397
2. Fahim S, Hasanien H, Turky R, Aleem SA, Calasan M (2022) A comprehensive review of models of photovoltaic modules and algorithms used in parameter extraction. Energies 15:8941. https://doi.org/10.3390/en15238941
3. Calasan M, Abdel-Aleem SHE, Zobaa AF (2021) A new approach for parameters estimation of double and triple diode models of photovoltaic cells based on iterative Lambert W function. Sol Energy 218:392–412. https://doi.org/10.1016/j.solener.2021.02.038
4. Vujosevic S, Calasan M, Micev M (2024) Hybrid walrus optimization algorithm techniques for optimized parameter estimation in single, double, and triple diode solar cell models. AIP Adv 14:085116. https://doi.org/10.1063/5.0223492
5. Ćalasan M, Vujošević S, Radonjić Mitić I (2025) Approximate analytical solutions for solar cell current-voltage characteristics: a four-diode model with two novel approaches. Energy Convers Manag 334:119835. https://doi.org/10.1016/j.enconman.2025.119835
6. Calasan M, Vujosevic S, Alruwaili M, Ahmed Ibrahim M (2025) Optimization of four-diode equivalent circuit models for solar cells: analytical formulation and performance enhancement. Alex Eng J 127:411–430. https://doi.org/10.1016/j.aej.2025.05.047
7. IEEE Standard for Floating-Point Arithmetic, IEEE Std 754™2008, IEEE Computer Society Sponsored by the Microprocessor Standards Committee (2008)
8. Roberts K, Valluri SR (2015) On calculating the current-voltage characteristic of multidiode models for organic solar cells. arXiv preprint arXiv:1601.02679
9. Corless RM, Jeffrey DJ (2002) The wright ω function. In: International conference on artificial intelligence and symbolic computation. Springer, pp 76–89
10. Ćalasan M (2025) Double-diode and triple-diode solar cell models: invertible approximate analytical expressions based on the g-function approach. J Comput Electron 24:18. https://doi.org/10.1007/s10825-024-02259-1
11. Rawa M, Calasan M, Abusorrah A, Alhussainy AA, Al-Turki Y, Ali ZM, Sindi H, Mekhilef S, Aleem SHEA, Bassi H (2022) Single diode solar cells—improved model and exact current–voltage analytical solution based on Lambert's W function. Sensors 22:4173. https://doi.org/10.3390/s22114173
12. Ćalasan M, Vujošević S (2025) Novel triple diode solar cells equivalent circuit models with Lambert W function expressions. IEEE J Electron Devices Soc 13:49–53. https://doi.org/10.1109/JEDS.2024.3523278
13. Micev M, Calasan M, Radulovic M (2024) Single diode solar cell models: review and comparison. ISGT Europe, Dubrovnik, Croatia. ISBN PRINT 978-953-184-297-6

14. Ćalasan M, Vujošević S, Micev M, Radonjić I, Petronijević M (2025) Enhanced two-diode model for solar cells: the impact of series resistance repositioning on performance and accuracy. In: Alternative energy sources, materials & technologies (AESMT'25), vol 7. Sofia, Bulgaria, pp 44–45. ISSN 2603 364X

15. Ali ZM, Ćalasan M, Mostafa MH, Abdel Aleem SHE (2024) Analytical modeling of novel equivalent circuits of double diode solar cell circuits using a special transcendental function approach. PLoS ONE 19(11):e0313713. https://doi.org/10.1371/journal.pone.0313713

16. Ćalasan M, Vujošević S, Bakić K (2025) Enhanced single-diode solar cell model: analytical solutions using Lambert W function and circuit innovations. IEEE J Electron Devices Soc 13:501–509

17. Calasan M, Vujosevic S, Micev M (2025) Novel single diode solar cell nonlinear model: optimization and validation. Comput Electr Eng 123:110150. https://doi.org/10.1016/j.compeleceng.2025.110150

18. Ćalasan M, Vujošević S, Krunić G (2025) Towards enhanced photovoltaic modeling: new single diode Model variants with nonlinear ideality factor dependence. Eng Sci Technol Int J 65:102037. https://doi.org/10.1016/j.jestch.2025.102037

19. Ćalasan M, Jovanović D, Rubežić V, Mujović S, Đukanović S (2019) Estimation of single-diode and two-diode solar cell parameters by using a chaotic optimization approach. Energies 12:4209. https://doi.org/10.3390/en12214209

20. Lun S, Wang S, Yang G, Guo T (2015) A new explicit double-diode modeling method based on Lambert W-function for photovoltaic arrays. Sol Energy 116:69–82. https://doi.org/10.1016/j.solener.2015.03.043

21. Ćalasan M, Vujošević S, Micev M, Alruwaili M, Wijaya AA (2024) Triple-diode solar cell current optimization—an analytical solution based on the Lambert W function. Alex Eng J 104:95–114. https://doi.org/10.1016/j.aej.2024.06.035

22. Ćalasan M (2025) Iterative solution of the current-voltage relationship in a four-diode solar cell model using the Lambert W equation. Sol Energy 290:113390. https://doi.org/10.1016/j.solener.2025.113390

Chapter 3
Application of Lambert W Function in Perovskite Solar Cell Modeling

This chapter consists of three parts. The first part provides basic information on perovskite solar cells. The second part describes the mathematical model of the perovskite solar cell, while the third part presents a numerical example.

3.1 Basic Information on Perovskite Solar Cells

From the standpoint of modern energy systems, solar energy is considered one of the most promising renewable energy sources due to its availability, sustainability, and the continuously decreasing costs of conversion technologies. Solar cells, also known as photovoltaic (PV) cells, directly convert sunlight into electricity through the photovoltaic effect. Within the various generations of solar cells—ranging from traditional silicon-based to more advanced organic and thin-film technologies—perovskite solar cells (PSC) have emerged as a revolutionary technology thanks to their rapid progress in efficiency and the simplicity of fabrication. In fact, PSCs represent a new generation of photovoltaic devices distinguished by their exceptionally high light-to-electricity conversion efficiency, which already exceeds 25% [1–4].

It can be stated that over the last decade, perovskite solar cells (PSCs) have established themselves as one of the most intensively studied research areas in the field of solar technology development. This status is owed to their unique combination of high efficiency, favorable optoelectronic properties, and compatibility with low-cost fabrication techniques. Unlike silicon cells, which require expensive and energy-intensive crystal processing, perovskites can be deposited using a variety of low-temperature methods, enabling the production of flexible and inexpensive solar modules.

The term "perovskite" originates from the specific crystal structure of the active layer in the cell, which is responsible for light absorption. These materials, most commonly based on hybrid organic-inorganic lead or tin halide compounds, exhibit

© The Author(s), under exclusive license to Springer Nature Switzerland AG 2026

M. Ćalasan and S. Vujošević, *Analytical Solutions of Nonlinear Power System Models Using the Lambert W Function*, SpringerBriefs in Electrical and Computer Engineering, https://doi.org/10.1007/978-3-032-09579-4_3

outstanding optoelectronic properties: strong light absorption, high charge-carrier mobility, tunable band gap, and low exciton binding energy. Such characteristics enable efficient charge separation and contribute to the high conversion of solar energy into electricity. Altogether, these features make PSCs exceptionally well-suited for photovoltaic applications.

The most significant advantages of perovskite solar cells are:

- High conversion efficiency—Laboratory-scale PSCs have achieved efficiencies exceeding 25%, while tandem configurations can reach up to 30%.
- Thin active material layers—Enable reduced material consumption without compromising performance.
- Tunable band gap—The absorption spectrum can be tailored by adjusting the material composition.
- Simple, low-temperature, and potentially low-cost fabrication—suitable for industrial applications.
- Flexibility and transparency—Allow the development of flexible and transparent solar modules suitable for various applications, including integration into building elements.
- Outstanding optoelectronic properties—including high carrier mobility, strong light absorption, and efficient charge separation.
- Potential for industrial-scale manufacturing—Techniques such as roll-to-roll processing and inkjet printing enable the transition from laboratory-scale proto-types to mass production.

One of the most commonly employed architectures of perovskite solar cells is the planar heterojunction structure, which consists of layers deposited in flat, parallel configurations. A typical planar structure includes:

- a transparent conductive oxide (TCO),
- an electron transport layer (ETL),
- a perovskite absorption layer,
- a hole transport layer (HTL), and
- a metallic contact electrode.

Its geometrical clarity and layered structure not only allow for simpler fabrication in laboratory and industrial environments but also provide a precise basis for studying fundamental processes, including carrier generation, separation, transport, and recombination. For this reason, this structure is widely used in the development of numerical models and in the optimization of device performance, playing a key role in the understanding and advancement of perovskite technologies.

The development of perovskite solar cells has demonstrated that their high efficiency and ease of fabrication can be successfully combined, paving the way for a new generation of photovoltaic technologies. However, the results obtained from experimental studies cannot fully explain the complex mechanisms governing the performance of these devices. Processes such as charge-carrier generation, separation, transport, and recombination within the perovskite layers, as well as the influence of contact materials and interfaces, require a detailed understanding and quantitative description.

Therefore, the development of mathematical models that incorporate these processes and enable performance prediction based on physical parameters is significant. Such models not only contribute to the fundamental understanding of perovskite solar cell operation but also serve as a practical tool for optimizing device structures and fabrication processes. In this way, modeling serves as a bridge between theoretical knowledge and practical realization, with its importance being especially evident in cases of new architectures and configurations, where experimental approaches alone are often insufficient to provide a comprehensive explanation of the observed effects.

3.2 Mathematical Model of Perovskite Solar Cell

To achieve a comprehensive understanding of the physical processes governing perovskite solar cell operation, it is essential to develop accurate mathematical models that link structural and material characteristics to the device's electrical response. Modeling a perovskite solar cell involves translating the key physical processes into a system of equations that describe its current–voltage characteristics. In practice, the equivalent circuit approach is most commonly employed, as it provides a concise and intuitive representation of the dominant transport and loss mechanisms, while simultaneously allowing direct comparison of theoretical predictions with experimental results.

3.2.1 Analytical Solution for PSC Current-Voltage Characteristics

One of the standard approaches in the theoretical modeling of perovskite solar cells is the two-diode equivalent circuit model (Fig. 3.1), which enables a quantitative analysis of the fundamental mechanisms of charge-carrier generation, separation, transport, and recombination. This model accounts for processes occurring both within the bulk of the active material and at the interfaces between the layers of the cell. Owing to its simplicity and analytical strength, this approach is widely used to correlate physical parameters with the electrical response of the device, representing a powerful tool for optimizing efficiency and reliability.

The equivalent electrical circuit of a perovskite solar cell consists of two series-connected diodes (D_1 and D_2), a series resistance R_S, a shunt resistance R_P, and a photocurrent source I_{PV}, which is generated as a result of light absorption in the active layer of the cell. The diodes model the two active junctions that are formed at the electron and hole transport layers due to light absorption in the cell. The

Fig. 3.1 PSC equivalent circuit

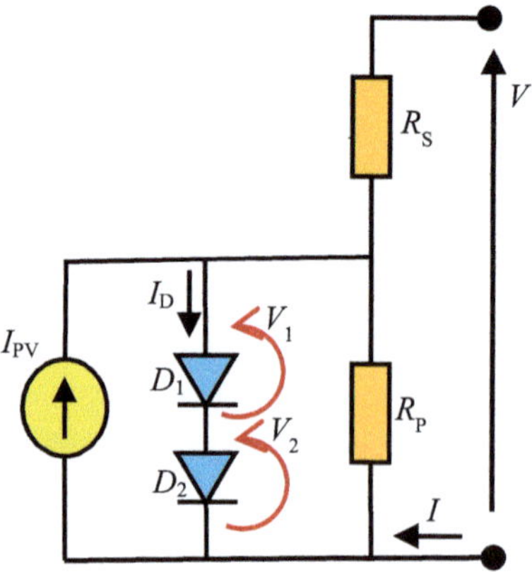

series resistance represents the internal losses associated with the flow of charge carriers, while the shunt resistance reflects the effect of reverse saturation current in the junctions.

From a mathematical standpoint, the current–voltage characteristic of a perovskite solar cell can be expressed as follows:

$$I = I_{PV} - I_D - \frac{V + IR_S}{R_P},\tag{3.1}$$

where I_D represents the current through diodes D_1 and D_2. The following expression can also be formulated:

$$I_D = I_{01}\left(e^{\frac{V_1}{a_1 V_{th}}} - 1\right) = I_{02}\left(e^{\frac{V_1}{a_2 V_{th}}} - 1\right),\tag{3.2}$$

where

- I_{PV}—photo-generated current, proportional to the illumination intensity [A],
- I_{01} and I_{02}—inverse saturation currents of the first and second diode [A],
- a_1 and a_2—ideality factors of the first and second diode,
- R_S—series resistance [Ω],
- R_P—shunt (parallel) resistance [Ω],
- V_{th} thermal voltage ($=k_BT/q$)[V],
- q—elementary charge (1.602×10^{-19} C),
- k_B—Boltzmann constant (1.38×10^{-23} J/K),
- T—absolute temperature [K].

Based on the physical characteristics of the material and in accordance with the available literature, the following expressions can be formulated:

$$I_{01} \propto P_n \frac{D_p}{L_p}$$

$$I_{02} \propto N_p \frac{D_n}{L_n}, \tag{3.3}$$

where D_n and D_p represent the diffusion coefficients of electrons and holes, L_n and L_p denote the diffusion lengths of electrons and holes. At the same time, P_n and N_p indicate the minority carrier concentrations. Furthermore, for these cells, the following approximation is adopted:

$$\frac{D_p}{L_p} \approx \frac{D_n}{L_n}. \tag{3.4}$$

When the carrier concentrations of electrons and holes are similar, as in the case of PSCs, the following relation holds:

$$I_{01} = I_{02}, \tag{3.5}$$

thus, the relation can be expressed as:

$$\frac{V_1}{n_1} = \frac{V_2}{n_2}, \tag{3.6}$$

$$V_1 + V_2 = V + IR_S, \tag{3.7}$$

and therefore

$$V_2 = \frac{a_2}{a_1 + a_2}(V + IR_S). \tag{3.8}$$

In conclusion, the functional dependence between current and voltage in the perovskite solar cell model is given by the following equation:

$$I = I_{PV} - I_0 \left(e^{\frac{V + IR_S}{(a_1 + a_2)V_{th}}} - 1 \right) - \frac{V + IR_S}{R_P}. \tag{3.9}$$

The presented formula represents the final mathematical expression that describes the current–voltage characteristic of the perovskite solar cell, based on the two-diode model with included series and parallel resistance.

Evidently, the current–voltage characteristic of the analyzed perovskite solar cell model represents a mathematical expression of pronounced nonlinearity. Accordingly, Eq. (3.9) can be reformulated in the following form [5, 6]:

$$I = \frac{R_P(I_{PV} + I_0) - V}{R_S + R_P} - V_{th}\left(\frac{a_1 + a_2}{R_3}\right)W(\alpha_I), \tag{3.10}$$

where

$$\alpha_{\mathrm{I}} = \frac{I_0 R_{\mathrm{P}} R_{\mathrm{S}}}{V_{\mathrm{th}}(a_1 + a_2)(R_{\mathrm{S}} + R_{\mathrm{P}})} \exp\left(\frac{R_{\mathrm{P}}(R_{\mathrm{S}} I_{\mathrm{PV}} + R_{\mathrm{S}} I_0 + V)}{V_{\mathrm{th}}(a_1 + a_2)(R_{\mathrm{S}} + R_{\mathrm{P}})}\right), \tag{3.11}$$

with W denoting the Lambert W function.

Furthermore, by applying the Lambert W function, it is also possible to express the voltage–current relationship, which takes the following form:

$$V = (I_{\mathrm{PV}} + I_0)R_{\mathrm{P}} - I(R_{\mathrm{S}} + R_{\mathrm{P}}) - V_{\mathrm{th}}(a_1 + a_2)W(\alpha_{\mathrm{V}}), \tag{3.12}$$

where

$$\alpha_{\mathrm{V}} = \frac{I_0 R_{\mathrm{P}}}{(a_1 + a_2)V_{\mathrm{th}}} \exp\left(\frac{R_{\mathrm{P}}(I_{\mathrm{PV}} + I_0 - I)}{V_{\mathrm{th}}(a_1 + a_2)}\right). \tag{3.13}$$

3.2.2 Analytical Solution for PSC Voltage-Current Characteristics

In the context of analytical modeling of the invertible current–voltage characteristics of perovskite solar cells, the traditional application of the Lambert W function in certain cases provides satisfactory solutions for expressions where the electric current appears as the unknown variable. However, it has been demonstrated that this approach is not universally applicable, particularly in operating regions corresponding to low or no load. Hence, it is evident that the Lambert W function is not suitable for the entire voltage range of PSC devices, and that only through the g-function a consistent and applicable analytical expression for the $V(I)$ dependence can be obtained.

On this basis, a g-function formulation is introduced, which not only incorporates the effects of series and parallel resistance but also enables an accurate description of cell behavior under boundary operating regimes. After the appropriate mathematical transformations, the voltage–current dependence can be expressed in the following form [7]:

$$V = (a_1 + a_2)g(\alpha_{\mathrm{I}}) - (a_1 + a_2)\log\left(\frac{I_0 R_{\mathrm{P}}}{(a_1 + a_2)V_{\mathrm{th}}}\right) - I R_{\mathrm{S}}, \tag{3.14}$$

where

$$\alpha_{\mathrm{I}} = \log\left(\frac{I_0 R_{\mathrm{P}}}{(a_1 + a_2)V_{\mathrm{th}}}\right) + \frac{R_{\mathrm{P}}(I_{\mathrm{PV}} + I_0 - I)}{(a_1 + a_2)V_{\mathrm{th}}}. \tag{3.15}$$

This approach represents an important contribution to the theoretical modeling of PSCs and proves to be an indispensable tool in the numerical simulation and optimization of these devices.

3.2.3 MATLAB Code for Inductor PSC Current Voltage Representation

The MATLAB implementation presented below provides the calculation of the PSC current–voltage and voltage–current characteristics.

```
count=0;
    for voltage=0.0:0.025:0.99
       count=count+1;
          I=I0*Rs*Rp/(a1a2*Vth*(Rs+Rp))*exp((Rp*(Rs*Ipv+Rs*I0+voltage))/
(a1a2*Vth*(Rs+Rp)));
       xIplot(count)=xI;
                                  Currentplot(count)=(Rp*(Ipv+I0)-voltage)/(Rs+Rp)-
a1a2*Vth*lambertw(xI)/Rs;
       Voltageplot(count)=voltage;
    end
    count=0;
    for current=0.0:0.0001:0.01864
       count=count+1;
       xU=I0*Rp/(a1a2*Vth)*exp((Rp/(a1a2*Vth)*(Ipv+I0-current)));
       betaUcrta(count)=xU;
                              VoltageWplot(count)=Rp*(Ipv+I0)-(Rp+Rs)*current-
a1a2*Vth*lambertw(xU);
       CurrentWplot(count)=current;
    end
     count=0;
    for current=0.0:0.00001:0.01865
       count=count+1;
       CurrentGplot(count)=current;
       alfa=log(I0*Rp/(a1a2*Vth))+((Rp*(Ipv+I0-current))/(a1a2*Vth));
       alfaIcount(count)=alfa;
         if alfa<-exp(1)
            y0=alfa;
         end
         if alfa>exp(1)
            y0=log(alfa);
         end
         if alfa<exp(1) && alfa>=-exp(1)
            y0=-exp(1)+(1+exp(1))/(2*exp(1))*(alfa+exp(1));
```

```
        end
        epsilon=10^-15;
        for counter=1:10000
                    y1=y0-(2*(y0+exp(y0)-alfa)*(1+exp(y0)))/(2*(1+exp(y0))^2-
(y0+exp(y0)-alfa)*exp(y0));
            if abs(y1-y0)<epsilon
              break
            end
            y0=y1;
          end
       gfunction=(y1);
                        VoltageGplot(count)=a1a2*Vth*gfunction-Rs*current-
a1a2*Vth*log((Rp*I0)/(a1a2*Vth));

end
```

3.3   Numerical Results

This section presents numerical results for the current–voltage (I–V) and voltage–current (V–I) characteristics calculated for the perovskite solar cell (PSC). To test the application of analytical current–voltage and voltage–current approaches, represented through the Lambert W function and the g-function, respectively, the following solar cell parameters were used:

$R_P = 5050;$

$R_S = 2.72;$

$I0 = 82.3*10^{-9};$

$I_{PV} = 0.01865;$

$a_1a_2 = 3.074;$

The corresponding current–voltage characteristics are presented in Fig. 3.2, where a clear agreement can be observed among all three modeling approaches applied to PSCs. However, for high voltage values, determining the voltage from the current using the Lambert W function is not possible.

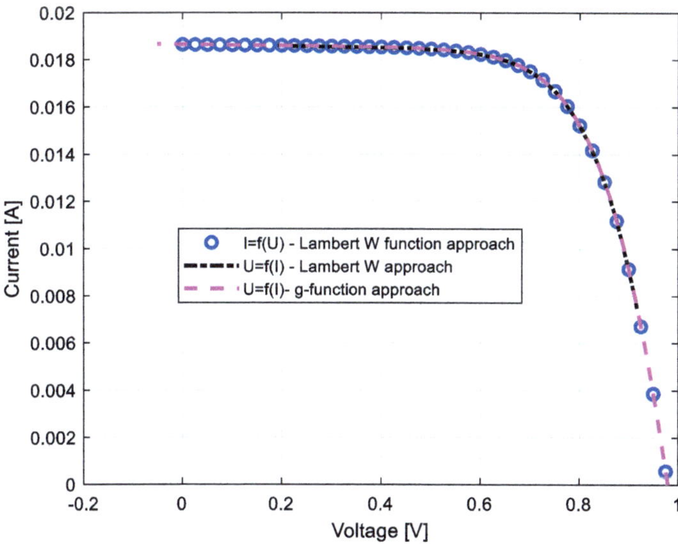

Fig. 3.2 PSC current-voltage characteristics

References

1. Krishna BG, Sundar Ghosh D, Tiwari S (2021) Progress in ambient air-processed perovskite solar cells: insights into processing techniques and stability assessment. Sol Energy 224:1369–1395. https://doi.org/10.1016/j.solener.2021.07.002
2. Huang Y, De SQ, Xu W, He Y, Yin WJ (2017) Halide perovskite materials for solar cells: a theoretical review. Wuli Huaxue Xuebao/Acta Phys Chim Sin 33:1730–1751. https://doi.org/10.3866/PKU.WHXB201705042
3. Tang S, Yan J, Chen L, Zhang W, Tan L (2024) Circuit modeling and analysis of hysteresis effect of perovskite photovoltaic cells. Sol Energy Mater Sol Cells 278:113182. https://doi.org/10.1016/j.solmat.2024.113182
4. Kabir MZ (2024) Analytical model for current-voltage characteristics in perovskite solar cells incorporating bulk and surface recombination. Micromachines 15(8):972. https://doi.org/10.3390/mi15080972
5. Rawa M, Al-Turki Y, Sindi H, Ćalasan M, Ali ZM, Abdel Aleem SHE (2023) Current-voltage curves of planar heterojunction perovskite solar cells—novel expressions based on Lambert W function and special trans function theory. J Adv Res 44:91–108. https://doi.org/10.1016/j.jare.2022.03.017
6. Singh NS, Kumar L, Sharma VK (2021) Solving the equivalent circuit of a planar heterojunction perovskite solar cell using Lambert W-function. Solid State Commun 337:114439. https://doi.org/10.1016/j.ssc.2021.114439
7. Calasan M (2024) Perovskite solar cells—novel modeling approaches for invertible current-voltage characteristics. Electr Eng 106:4903–4912. https://doi.org/10.1007/s00202-024-02248-4

Chapter 4
Modeling the No-Load Starting Process of Induction Machines via Lambert W Function

This chapter is divided into several parts. The first part provides a brief description of the induction machine from the standpoint of its significance. The second part addresses the problem of starting an induction machine. The third part presents the analytical formulation of the time-speed and speed-time relationships during direct start-up. The fourth part represents the induction machine acceleration characteristics. In the last part, numerical examples are provided.

4.1 Basic Information About Induction Machines

An induction machine, also known as an asynchronous machine, is the most widely used type of electrical machine in modern power and industrial systems [1]. Due to its simplicity, reliability, and wide power range (from tens of watts to several megawatts), the induction machine is commonly employed in the drives of pumps, fans, compressors, conveyor belts, cranes, and increasingly in renewable energy systems and electric vehicles.

The wide range of applications of induction machines stems from several design and operational advantages that make them suitable for various technical requirements and working conditions. Key advantages include:

- Robust construction and high mechanical durability, especially in squirrel-cage rotors that do not contain brushes or slip rings, significantly reducing wear and maintenance requirements;
- Low operating costs, thanks to their simple mechanical structure and reliable technology, resulting in reduced service interventions and long-term operational expenses;
- High reliability, particularly in prolonged operation under variable and harsh industrial environments;

M. Ćalasan and S. Vujošević, *Analytical Solutions of Nonlinear Power System Models Using the Lambert W Function*, SpringerBriefs in Electrical and Computer Engineering, https://doi.org/10.1007/978-3-032-09579-4_4

- Energy efficiency, which in modern models meets international standards (IE2, IE3, IE4), enabling optimized electricity consumption;
- Flexibility in starting methods, allowing for direct-on-line starting as well as reduced voltage starting using star–delta connections, autotransformers, or soft starters, depending on system inertia and load requirements;
- Capability for precise speed and torque control, especially when combined with variable frequency drives (VFDs), enabling operation in variable regimes and integration into modern automated systems.

Thanks to these features, induction machines are today considered the standard solution in a wide range of industrial and infrastructural applications. Their simple construction, technological maturity, and high reliability make them suitable not only for traditional manufacturing processes but also for new areas of application in energy-efficient and sustainable technologies. Therefore, induction machines are increasingly being used not only in standard industrial plants but also in demanding applications such as wind turbine drives, pumped-storage hydroelectric plants, and electric vehicle propulsion systems.

4.2 Induction Machine Starting

A specific technical and scientific challenge associated with induction machines is the problem of the starting process. The starting process is particularly interesting given that transient phenomena during this phase can significantly affect both the machine itself and the power system to which it is connected [2–4].

In industrial practice, various starting methods are applied, the most notable being:

- **Direct online starting,**
- **Reduced voltage starting** (via star–delta connections or autotransformers), and
- **Start using soft starters or variable frequency drives (VFDs).**

Reduced voltage starting, regardless of the method used, enables the machine to start with current values that are typically close to the rated value. As a result, these starting methods are characterized by slower transient processes, since the machine requires more time to reach steady-state operation. However, when soft starters—power electronic devices—are used, higher-order harmonics are also generated.

Direct-on-line starting, although the simplest from both technical and design perspectives, can result in extremely high inrush currents that greatly exceed rated values. These currents typically reach 5 to 6 times the rated current of the machine, potentially causing thermal stress on stator windings, voltage sags in the local grid, and unwanted tripping of protection devices. These voltage sags do not affect only the machine itself but also all other consumers supplied from the same feeder or transformer substation.

In this context, understanding the dynamic behavior of the machine during startup is essential for accurately estimating the acceleration time, designing appropriate protection, and optimizing the commissioning process.

The no-load starting time of an induction machine represents an important indicator of the system's dynamic response and holds multiple implications in engineering practice. Firstly, it defines the upper limit of startup duration under real load conditions, serving as a reference value when dimensioning protective and control devices (e.g., time relays, thermal protections). Additionally, knowledge of this time allows for accurate estimation of rotor acceleration, thereby facilitating the analysis of transient effects in the network and the optimization of the overall startup process.

Consequently, the development of analytical expressions describing the speed–time relationship (and vice versa) becomes a valuable tool for understanding and evaluating the engineering performance of induction machines during no-load starting. Such models are essential for assessing acceleration time, analyzing transient effects, and optimizing the startup process, especially under no-load conditions.

For these reasons, the following sections will analyze the process of direct online starting of induction machines under no-load conditions, aiming to develop an analytical speed–time model based on the rational Kloss torque characteristic and its transformation using the Lambert W function.

4.3 Induction Machine Speed-Time and Time-Speed Curve During No-Load Direct Start-Up

The following differential equation fundamentally describes the mechanical behavior of an electrical machine:

$$J\frac{d\omega}{dt} = \sum M, \tag{4.1}$$

where J is the moment of inertia of the rotating system, ω denotes the angular velocity of the rotor, and M represents the total torque acting on the machine.

This fundamental relation can be further expressed, depending on the specific operating conditions, in the following form:

$$J\frac{d\omega}{dt} = M_{em} - M_L, \tag{4.2}$$

where M_{em} represents the electromagnetic torque developed by the machine, and M_L denotes the torque due to external mechanical load.

In the case of no-load start-up and neglecting friction losses in the bearings, the previous relation simplifies to:

$$J\frac{d\omega}{dt} = M_{em}. \tag{4.3}$$

The expression for the machine torque can be written either using the Kloss equation or by applying Thévenin's theorem. Below are the formulas for both approaches, presented side by side.

Kloss equation	Thevenin's theorem
$M = \dfrac{2M_{br}}{\frac{s}{s_{br}} + \frac{s_{br}}{s}}$	$M_{br} = \dfrac{3\left(V_T/\sqrt{3}\right)^2}{w_s} \dfrac{\frac{R_2}{s}}{\left(R_T + \frac{R_2}{s}\right)^2 + (X_T + X_2)^2}$
$t = J \int_{\omega_0}^{\omega_f} \dfrac{d\omega}{M}$	
$t = \dfrac{J\omega_s}{2M_{br}} \int_s^1 \left(\dfrac{s}{s_{br}} + \dfrac{s_{br}}{s}\right) ds$	$t = J \int_{\omega_0}^{\omega_f} \dfrac{d\omega}{M}$
$t = \dfrac{J\omega_s}{2M_{br}} \left(\dfrac{1-s^2}{2s_{br}} - s_{br}\ln(s)\right)$	$t = -\omega_s J \int_1^s \dfrac{ds}{\frac{V_T^2}{\omega_s} \frac{R_2/s}{(R_T + R_2/s)^2 + (X_T + X_2)^2}}$
	$t = \dfrac{J\omega_s^2}{V_T^2 R_2} \left(\dfrac{R_T^2 + (X_T + X_2)^2}{2}(1 - s^2) + 2R_T R_2(1 - s) + R_2^2\ln(s)\right)$

In the previous equation, we can use the following expressions for breaking torque and breaking slip:

$$M_{br} = \frac{3\left(V_T/\sqrt{3}\right)^2}{w_s} \frac{(X_T + X_2)}{(R_T + X_T + X_2)^2 + (X_T + X_2)^2} \tag{4.4}$$

$$s_{br} = \frac{R_2}{\sqrt{(R_T)^2 + (X_T + X_2)^2}},$$

where Thevenin's voltage is as follows

$$V_T = \frac{X_m}{(X_1 + X_m)} V, \tag{4.5}$$

while Thevenin resistance and reactance is as follows:

$$R_T = \frac{R_1 X_m^2}{R_1^2 + (X_1 + X_m)^2}, \tag{4.6}$$

$$X_T = \frac{R_1 X_m^2 + X_1^2 X_m + X_1 X_m^2}{R_1^2 + (X_1 + X_m)^2}, \tag{4.7}$$

and synchronous speed is as follows:

$$\omega_s = \frac{2\pi n_s}{60}. \tag{4.8}$$

In the previous equations, R_1 represents the stator resistance, while R_2 denotes the rotor resistance referred to the stator side. The reactances X_1 and X_2 correspond to the stator and rotor (also referred to as the stator side), respectively, whereas X_m

stands for the magnetizing reactance. Furthermore, n_s is the synchronous speed of the rotating magnetic field, and V represents the supply voltage applied to the machine.

4.3.1 Analytical Solution for Induction Machine Speed-Time Characteristics During Direct Startup

As can be seen, regardless of whether Kloss's approach is used or the Thevenin equivalent is applied, the expression for time as a function of speed results in a complex form that depends on both linear terms of slip and logarithmic components.

In this monograph, to demonstrate the application of the Lambert W equation, the slip–time relationship will be considered based on the Kloss expression.

Specifically, the equation for time:

$$t = \frac{J\omega_s}{2M_{br}}\left(\frac{1-s^2}{2s_{br}} - s_{br}\ln(s)\right),\tag{4.9}$$

can be transformed into the following form:

$$\frac{1}{s_{br}^2}s^2 = \frac{1}{s_{br}^2}e^{A(\tau-t)}e^{-\frac{1}{s_{br}^2}s^2},\tag{4.10}$$

where

$$\tau = \frac{J\omega_s}{4s_{br}M_{br}},\; A = \frac{4M_{br}}{J\omega_s s_{br}},\tag{4.11}$$

or more precisely into the form:

$$W = xe^{-W},\tag{4.12}$$

where W is a solution of the Lambert W function

$$W = \frac{s^2}{s_{br}^2},\tag{4.13}$$

and

$$x = \frac{1}{s_{br}^2}e^{A(\tau-t)}.\tag{4.14}$$

Accordingly, the expression for slip s takes the following form:

$$s = s_{\text{br}} \sqrt{W\left(\frac{1}{s_{\text{br}}^2} e^{A(\tau - t)}\right)}. \tag{4.15}$$

Based on the definition of slip, the rotor speed during startup can be expressed as:

$$n = n_{\text{s}}\left(1 - s_{\text{br}}\sqrt{W\left(\frac{1}{s_{\text{br}}^2} e^{A(\tau - t)}\right)}\right). \tag{4.16}$$

It should be emphasized that the derived expressions represent closed-form analytical solutions for the variation of slip and rotor speed during the startup process under no-load conditions. Unlike standard numerical procedures that require solving the differential equation of motion, this approach enables direct evaluation of the system's dynamics at any given moment in time.

Such a formulation significantly simplifies the analysis, reduces computational complexity, and makes the solutions suitable for use in optimization algorithms, simulation platforms, and automatic control systems.

In the literature, certain approaches have been proposed for solving the speed–time curve of an induction machine during direct start in a closed-form analytical manner. These approaches are based on the application of Taylor series expansion and the theory of Special Tran Functions Theory (STFT—Special Tran Function Theory).

If a Taylor series expansion is used, the rotor speed during startup can be approximated by the following relation:

$$n = n_{\text{s}}\left(1 - s_{\text{br}}\sqrt{\sum_{k=1}^{M} \frac{(-k)^{k-1}}{k!} \left(\frac{1}{s_{\text{br}}^2} e^{A(\tau - t)}\right)^k}\right). \tag{4.17}$$

On the other hand, when the STFT method is applied, the speed–time curve of the induction machine during direct start takes the following form:

$$n = n_{\text{s}}\left(1 - \sqrt{\frac{1}{s_{\text{br}}^2} e^{A(\tau - t)} \cdot \frac{\sum_{k=0}^{M} \frac{\left(\frac{1}{s_{\text{br}}^2} e^{A(\tau - t)}\right)^k (M-k)^k}{k!}}{\sum_{k=0}^{M+1} \frac{\left(\frac{1}{s_{\text{br}}^2} e^{A(\tau - t)}\right)^k (M+1-k)^k}{k!}}}\right). \tag{4.18}$$

Both Eqs. (4.17) and (4.18) represent original analytical solutions for calculating the speed of an induction machine during direct starting. Due to their simplicity and ease of implementation in software tools, such expressions enable direct computation of the instantaneous speed when machine parameters are known.

4.3.2   *MATLAB Code for Time-Speed and Speed-Time Calculation*

The MATLAB code below enables the determination of the induction machine time-speed and time-speed characteristics during direct start-up.

```
clc

clear all

% define machine parameters

VT=Xm/(X1+Xm)*V

RT=R1*Xm^2/( R1^2+(X1+Xm)^2)

XT=(Xm*R1^2+X1^2*Xm+X1*Xm^2)/(R1^2+(X1+Xm)^2)

Mbr=3/ws*(VT/sqrt(3))^2*(XT+X2)/((RT+XT+X2)^2+(XT+X2)^2)

sbr=R2/(sqrt(RT^2+(XT+X2)^2))

% Code

constA=3/ws*(VT/sqrt(3))^2*(R2);

constC=(XT+X2)^2;

constIM=J*(2*pi*ns/60)/(2*Mbr)

ds=0.0001

counter=0;

for x=0:ds:0.99999
    counter=counter+1;
    slip(counter)=x;
    speed(counter)=(1-x)*ws*60/2/pi;
                timeThevenin(counter)=J*ws/constA*((RT^2+(XT+X2)^2)/2*(1-
x^2)+2*RT*R2*(1-x)-R2^2*log(x) );
    timeKloss(counter)=constIM*((1-x^2)/(2*sbr)-sbr*log(x));

end

figure(1)

plot(timeThevenin,speed,'b','Linewidth',2)

hold on

plot(timeKloss,speed,'r--','Linewidth',2)

xlabel('Time [s]')

ylabel('Speed [r/min]')
```

```
legend('Thevenin','Kloss')

grid on

box on

constINV=4*Mbr/(J*ws*sbr);

thau=J*ws/(4*sbr*Mbr);

tstep=0.02;

tfinal=1;

counter=0;

for time=0:tstep:tfinal
    counter=counter+1;
        speedinv(counter)=ns*(1-sbr*sqrt(lambertw(1/sbr^2*exp(constINV*(thau-
time)))));
    timeinv(counter) = time;

end

figure(1)

hold on

plot(timeinv,speedinv,'kx','Linewidth',2)

hold on

legend('Thevenin','Kloss','Inverse Kloss')
```

4.4   Determining the Acceleration of an Induction Motor Using the Kloss Equation—Analytical Solution

In the previous analysis, the focus was on examining the dependence of speed on time during the no-load start-up of induction machines. However, for dynamic analysis and the assessment of mechanical stresses, it is often essential to also understand the acceleration as a function of time.

Starting from the fundamental relation for the electromagnetic torque of the induction machine, the rotor acceleration can be defined as:

$$\alpha = \frac{\mathrm{d}\omega}{\mathrm{d}t},\tag{4.19}$$

or as follows:

$$\alpha = \frac{d\omega}{dt} = \frac{2M_{br}}{J} \frac{\sqrt{W\left(\frac{1}{s_{br}^2}e^{A(\tau-t)}\right)}}{W\left(\frac{1}{s_{br}^2}e^{A(\tau-t)}\right)+1}. \tag{4.20}$$

By differentiating the previously derived expression, the maximum acceleration and the time at which it occurs can be determined. The expressions for these values are as follows [4]:

$$t = \tau - \frac{1+2\ln(s_{br})}{A}, \tag{4.21}$$

and

$$\alpha_{\max} = \frac{M_{br}}{J}. \tag{4.22}$$

MATLAB code for the determination of induction machine acceleration during direct start-up is given below.

```
constINV=4*Mbr/(J*ws*sbr);

thau=J*ws/(4*sbr*Mbr);

counter=0;

for time=0:tstep:tfinal
    counter=counter+1;
        speedinv(counter)=ns*(1-sbr*sqrt(lambertw(1/sbr^2*exp(constINV*(thau-time)))));
    ARG(counter)=1/sbr^2*exp(constINV*(thau-time));
                ACCEL(counter)=2*Mbr/J*(sqrt(lambertw(ARG(counter))))/
(lambertw(ARG(counter))+1);
    timeinv(counter) = time;
end

time_max_acc=thau-(1+2*log(sbr))/constINV

ACCEL_max=Mbr/J
```

4.5 Numerical Results

To demonstrate the applicability of the derived solutions, an induction machine with the following parameters was considered.

```
% Voltage, Inertia
```

$V = 460.$

$J = 1.662;$

% basic parameters

$ns = 1800;$

$ws = 2*pi*ns/60;$

$Ub = 460.$

$Sb = 37300.$

$Zb = Ub^2/Sb.$

$R1 = 0.087.$

$R2 = 0.228.$

$X1 = 0.302.$

$X2 = 0.302.$

$Xm = 13.08.$

It was assumed that the machine is directly started with a nominal voltage of 400 V. Figure 4.1 shows the numerical results obtained using the time-speed formulations (based on both the Kloss model and Thevenin's equivalent), as well as the inverse approach for the speed-time calculation. The two approaches based on the Kloss model (time-speed and speed-time) are consistent. In this case, the moment at which the maximum acceleration occurs is 0.3013 s, while the maximum value of the acceleration is 469.9072 rad/s^2.

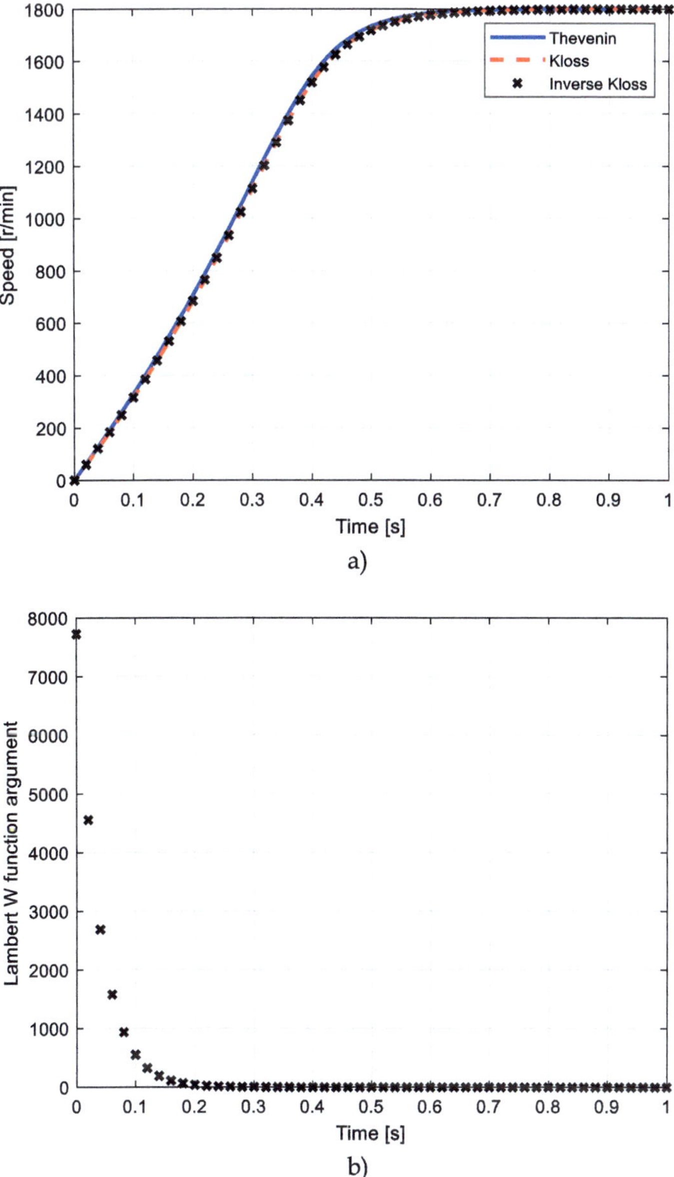

Fig. 4.1 Induction machine **a** speed-time characteristics, **b** Lambert W function argument change versus time, and **c** acceleration-time characteristics (rated voltage)

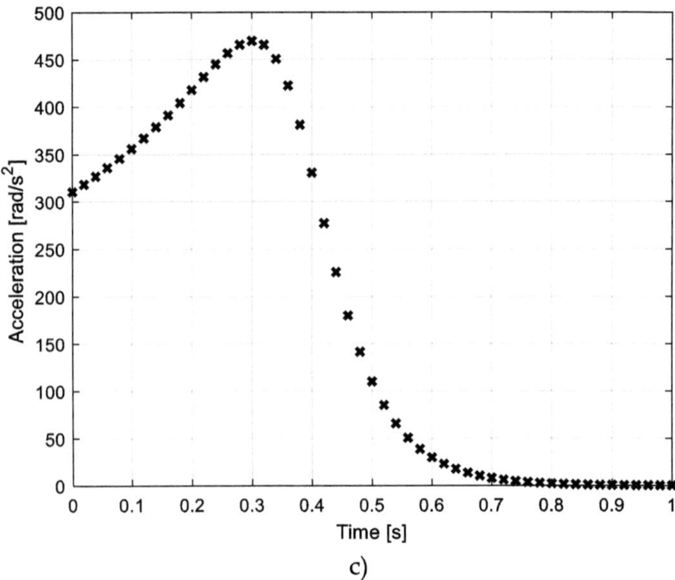

c)

Fig. 4.1 (continued)

References

1. Krause PC, Wasynczuk O, Sudhoff SD (2002) Analysis of electric machinery and drive system. Wiley, IEEE Press, New York
2. Ćalasan MP (2020) An invertible dependence of the speed and time of the induction machine during no-load direct start-up. Automatika 61(1):141–149. https://doi.org/10.1080/00051144. 2019.1689725
3. Ćalasan MP (2019) Analytical solution for no-load induction machine speed calculation during direct start-up. Int Trans Electr Energy Syst 29(4):e2777. https://doi.org/10.1002/etep.2777
4. Koljčević N, Fuštić Ž, Ćalasan M (2020) Analytical solution for determination of induction machine acceleration based on Kloss equation. Serbian J Electr Eng 17(2):247–256. https://doi. org/10.2298/SJEE2002247K

Chapter 5
Modeling the Mann Model of Proton Exchange Membrane Fuel Cells (PEMFC) Using Lambert W Function

This chapter is divided into several parts. The first part provides basic information about Proton Exchange Membrane Fuel Cells (PEMFCs). The second part discusses the mathematical modeling of PEMFCs. The third part presents the iterative Lambert W function procedure for PEMFC current calculation, along with key excerpts from the released MATLAB codes for all approaches. The final part of this chapter presents the numerical results.

5.1 Basic Information About PEMFC

The idea of clean and efficient energy generation has captivated the scientific and engineering community for decades. While renewable sources such as solar and wind power are often portrayed as ideal solutions, real-world experience reveals that these technologies are not without significant limitations. Their dependence on weather conditions, time of day, and geographic location—as well as challenges related to efficiency, production stability, and spatial requirements—demonstrates that sustainability alone does not ensure the overall reliability of the power system.

In this context, increasing attention is being directed toward technologies capable of complementing or stabilizing the operation of renewable energy sources, while also enabling reliable and efficient electricity generation regardless of external conditions. Among these, proton exchange membrane fuel cells (PEMFCs) stand out due to their combination of high efficiency, environmental compatibility, modularity, and fast dynamic response [1].

Fuel cells are electrochemical devices that enable the direct conversion of the chemical energy of a fuel into electrical energy, without intermediate conversion into mechanical energy, which significantly enhances overall system efficiency. In a typical proton exchange membrane fuel cell (PEMFC), hydrogen-rich fuel enters the anode, where protons and electrons are separated. The protons pass through the electrolyte membrane, while the electrons flow through an external circuit, generating electric current. At the cathode, protons and electrons recombine with oxygen, producing water and heat as the only byproducts.

As a result of this process, PEMFCs are characterized by exceptionally high efficiency, which can exceed 90% in cogeneration mode (simultaneous production of electricity and heat). In addition, fuel cell operation is silent, vibration-free, and does not produce harmful gas emissions, making them suitable for deployment in urban environments, in proximity to end users. This eliminates the need for complex and energy-loss-prone distribution networks.

Another important characteristic of PEM fuel cells is their rapid responsiveness to changes in power demand, which makes them particularly suitable for systems with pronounced load variability. An additional advantage of PEMFC technology lies in its ability to quickly modulate output power, making it highly appropriate for applications that require both flexibility and stable energy response.

Furthermore, these cells feature compact dimensions, low operating temperatures (60–80 °C), rapid start-up, and the potential for miniaturization. These properties enable their integration across a wide range of applications—from stationary power plants and hydrogen-powered vehicles to portable electronic devices.

Considering all these advantages, it is not surprising that PEM fuel cells have become the subject of intensive research efforts worldwide, both within the academic community and in industrial practice. Enhancing their performance and reliability requires the development of accurate mathematical models that can capture their behavior under various operating conditions. For optimized design, control, and integration into larger systems, it is essential to formulate robust modeling frameworks that can reliably represent their operation in real-world environments.

Such models often rely on analogies with electrical circuits, allowing the simulation of key output characteristics of the fuel cell, including current–voltage behavior, output power, efficiency, thermal stability, and dynamic system response. Among the various modeling approaches, the Mann model holds a particularly prominent position, as it is widely used to analyze the current–voltage relationship in PEMFC systems. These types of models provide deeper insight into the internal processes of the fuel cell and form the foundation for the development of advanced control algorithms and system performance prediction.

5.2 Mathematical Model of PEMFCs

Modeling of proton exchange membrane fuel cells (PEMFCs) is essential for understanding and predicting their behavior under varying operating conditions. In this context, the well-established Mann's model is employed, as it has proven to be an effective tool for accurately describing the polarization characteristics, that is, the relationship between voltage and current in PEM fuel cells.

Considering that the output voltage of a single fuel cell typically ranges from 0.9 V to 1.1 [V], achieving higher voltages or greater total current requires the connection of multiple cells. This is accomplished through series, parallel, or combined (series-parallel) configurations. Series connections increase the overall voltage of the system, while parallel connections enhance its current-carrying capacity. In this way, the parameters of the complete PEMFC system can be tailored to meet the specific requirements of the intended application, whether in mobile devices, industrial facilities, or power grid systems.

The output voltage of the stack can be represented by the following equation [2–4]

$$V_{\text{stack}} = N_{\text{cells}}(E_{\text{Nernst}} - V_{\text{act}} - V_{\Omega} - V_{\text{conc}}), \tag{5.1}$$

where N_{cells}, E_{Nernst}, V_{act}, V_{Ω} and V_{conc} denote the number of series-connected fuel cells, the Nernst voltage of a single cell, the activation overpotential, the ohmic voltage drop per cell, and the concentration overpotential, respectively.

The Nernst voltage, E_{Nernst}, can be described by the following relation:

$$E_{\text{Nernst}} = 1.229 - 0.85 \times 10^{-3}(T_{\text{fc}} - 298.15)$$
$$+ 4.3085 \times 10^{-5} \times T_{\text{fc}} \times \log\left(P_{\text{H}_2}\sqrt{P_{\text{O}_2}}\right), \tag{5.2}$$

where T_{fc} is the operating temperature of the cell [K], and P_{H_2} and P_{O_2} are the partial pressures of hydrogen and oxygen [atm], respectively.

The mathematical expression for the activation overpotential, V_{act}, is given by:

$$V_{\text{act}} = -\left[\xi_1 + \xi_2 T_{\text{fc}} + \xi_3 T_{\text{fc}} \log(C_{\text{O}_2}) + \xi_4 T_{\text{fc}} \log(I_{\text{fc}})\right]; \tag{5.3}$$

$$C_{\text{O}_2} = \left(\frac{P_{\text{O}_2}}{5.08 \cdot 10^6}\right)e^{\frac{498}{T_{\text{fc}}}},$$

where I_{fc} is the fuel cell current, and $\xi_1 - \xi_4$ represent model constants.

The ohmic voltage drop per cell is calculated using the following relation:

$$V_{\Omega} = I_{\text{fc}}(R_{\text{m}} + R_{\text{c}}), \tag{5.4}$$

where R_{m} denotes the membrane resistance and R_{c} the contact resistance, both expressed in ohms. The membrane resistance is given by the following expression:

$$R_{\text{m}} = \frac{\rho_{\text{m}} l}{M_{\text{A}}}; \tag{5.5}$$

where

$$\rho_{\mathrm{m}} = \frac{181.6\left[1 + 0.03\left(\frac{I_{\mathrm{fc}}}{M_{\mathrm{A}}}\right) + 0.062\left(\frac{T_{\mathrm{fc}}}{303}\right)^{2}\left(\frac{I_{\mathrm{fc}}}{M_{\mathrm{A}}}\right)^{2.5}\right]}{\left(\lambda - 0.634 - 3\frac{I_{\mathrm{fc}}}{M_{\mathrm{A}}}\right)e^{4.18\cdot\frac{T_{\mathrm{fc}}-303}{T_{\mathrm{fc}}}}}; \tag{5.6}$$

where ρ_{m} represents the membrane resistivity [Ω.cm], l is the membrane thickness [cm], M_{A} is the membrane area [cm^2], and λ is an adjustable parameter.

The concentration overpotential, V_{conc}, can be expressed by the following relation:

$$V_{\mathrm{conc}} = -\beta \log\left(1 - \frac{I}{I_{\mathrm{max}}}\right), \tag{5.7}$$

where β is a parametric coefficient, I represents the instantaneous current [A], and I_{max} corresponds to the maximum current value [A].

The formulated set of equations provides the foundation for describing the operation of PEM fuel cells, enabling the analysis of their performance and behavior under various operating conditions.

5.3   Lambert W Function Iterative Procedure for PEMFC Current Calculation

This section consists of two parts. The first part describes the mathematical background for modeling the current–voltage characteristics of PEMFCs. The second part presents the implemented MATLAB codes for these calculations.

5.3.1   Mathematical Background

In addition to the relation where voltage is expressed as a function of current, it is important to note that when measuring the current–voltage characteristics of PEM fuel cells using a variable load, the inverse characteristics are essentially obtained. The measurement procedure is as follows: a specific current value is set, and the corresponding voltage across the cell is measured. Then, the load resistance is changed, resulting in a new current value, and the corresponding voltage is measured again. In this way, the fuel cell's behavior changes with each measurement, since the operating conditions are dynamically altered, and each new load value produces a new current–voltage pair [5–7].

Within this context, the inverse relationship—where current is expressed as a function of voltage—is also considered. This formulation has often been overlooked in the existing literature and has not been the subject of detailed analysis. Nevertheless, it holds particular importance in both experimental and simulation studies, as it enables the direct determination of the PEM fuel cell current based on a known (measured) voltage value.

In this case, the cell current is determined using the following procedure:

$$\Theta \cdot \exp(\Theta) = H, \tag{5.8}$$

where $H = \dfrac{a_4}{b_1 g}\left(\dfrac{(1-m\cdot I_{\mathrm{fc}})^{\beta\left(\frac{g-hI_{\mathrm{fc}}}{I_{\mathrm{fc}}}\right)}\cdot I_{\mathrm{fc}}^{-bhI_{\mathrm{fc}}}}{e^{a_3+a_5 I_{\mathrm{fc}}^2+a_6 I_{\mathrm{fc}}^{3.5}}} \right)^{\frac{1}{b_1 g}}$ and

$$\Theta = \frac{a_4}{b_1 g} I_{\mathrm{fc}}. \tag{5.9}$$

In expressions (5.8) and (5.9), the variables take the following form:

$$a = \xi_1 + \xi_2 T_{\mathrm{fc}} + \xi_3 T_{\mathrm{fc}} \log\left(C_{O_2}\right)$$

$$b = \xi_4 \cdot T_{\mathrm{fc}}$$

$$c = R_{\mathrm{c}}$$

$$d = \frac{1}{M_{\mathrm{A}}} \cdot 181.6$$

$$ee = \frac{0.03}{M_{\mathrm{A}}}$$

$$f = 0.062 \left(\frac{T_{\mathrm{fc}}}{303}\right)^2 \left(\frac{1}{M_{\mathrm{A}}}\right)^{2.5}$$

$$g = (\lambda - 0.634)e^{\frac{4.18(T_{\mathrm{fc}}-303)}{T_{\mathrm{fc}}}}$$

$$h = \frac{3}{M_{\mathrm{A}}} e^{\frac{4.18(T_{\mathrm{fc}}-303)}{T_{\mathrm{fc}}}}$$

$$m = \frac{1}{M_{\mathrm{A}} \cdot J_{\max}}$$

$$b_1 = -b$$

$$a_2 = \frac{V_{\mathrm{stack}}}{N_{\mathrm{cells}}} - E_{\mathrm{Nernst}} - \left(\xi_1 + \xi_2 T_{\mathrm{fc}} + \xi_3 T_{\mathrm{fc}} \log\left(C_{O_2}\right)\right)$$

$$a_3 = a_2 g = \left(\frac{V_{\text{stack}}}{N_{\text{cells}}} - E_{\text{Nernst}} - a \right)(\lambda - 0.634)e^{\frac{4.18(T_{\text{fc}}-303)}{T_{\text{fc}}}}$$

$$a_4 = -a_2 h + cg + d$$

$$= R_{\text{c}}(\lambda - 0.634)e^{\frac{4.18(T_{\text{fc}}-303)}{T_{\text{fc}}}} + \frac{181.6}{M_A}$$

$$- \frac{3e^{\frac{4.18(T_{\text{fc}}-303)}{T_{\text{fc}}}}}{M_A}\left(\frac{V_{\text{stack}}}{N_{\text{cells}}} - E_{\text{Nernst}} - \left(\xi_1 + \xi_2 T_{\text{fc}} + \xi_3 T_{\text{fc}} \log(C_{O_2}) \right) \right)$$

$$a_5 = -c \cdot h + d \cdot ee = \frac{181.6}{M_A}\left(\frac{0.03}{M_A} \right) - \frac{3R_{\text{c}}}{M_A}e^{\frac{4.18(T_{\text{fc}}-303)}{T_{\text{fc}}}}$$

$$a_6 = d \cdot f = 0.062\left(\frac{181.6}{M_A} \right)\left(\frac{T_{\text{fc}}}{303} \right)^2 \left(\frac{1}{M_A} \right)^{2.5}$$

It is evident that both Θ and H depend on the fuel cell current and can therefore be treated as functions of I_{fc}. In this context, an iterative procedure is formulated to determine the current value based on a known voltage of the PEM fuel cell.

The iteration procedure for solving the current-voltage expression begins by defining an initial approximation for the current, denoted as $I_{\text{fc}(0)}$. Based on this initial value, the corresponding initial value of the parameter H, denoted $H_{(0)}$, is computed using the appropriate functional relation derived from the mathematical model.

The resulting function H depends on the instantaneous value of the current and is defined as follows:

$$H_{(0)} = \frac{a_4}{b_1 g}\left(\frac{\left(1 - m \cdot I_{\text{fc}(0)}\right)^{\beta\left(\frac{g - h I_{\text{fc}(0)}}{I_{\text{fc}}} \right)} \cdot I_{\text{fc}(0)}^{-bh I_{\text{fc}}}}{e^{a_3 + a_5 I_{\text{fc}(0)}^2 + a_6 I_{\text{fc}(0)}^{3.5}}} \right)^{\frac{1}{b_1 g}}. \tag{5.10}$$

In the same iteration, the value of the parameter Θ is updated according to the following relation:

$$H_{(0)} = \Theta_{(1)} \cdot \exp\left(\Theta_{(1)}\right). \tag{5.11}$$

Following the previously defined function H, the solution to Eq. (5.11) can be explicitly expressed using the Lambert W function:

$$\Theta_{(1)} = W\left(H_{(0)}\right), \tag{5.12}$$

where W denotes the solution of the Lambert W equation.

After that, the updated value of H, denoted as $H_{(1)}$, can be determined as:

$$H_{(1)} = \frac{a_4}{b_1 g} \left(\frac{\left(1 - m \cdot I_{\text{fc}(1)}\right)^{\beta\left(\frac{g - h I_{\text{fc}(1)}}{I_{\text{fc}}}\right)} \cdot I_{\text{fc}(1)}^{-bh I_{\text{fc}}}}{e^{a_3 + a_5 I_{\text{fc}(1)}^2 + a_6 I_{\text{fc}(1)}^{3.5}}} \right)^{\frac{1}{b_1 g}}, \tag{5.13}$$

where

$$I_{\text{fc}(1)} \frac{b_1 g}{a_4} = \Theta_{(1)}. \tag{5.14}$$

The iterative procedure continues until the value of variable H in the k-th and $(k + 1)$-th iterations approaches a predefined value.

Finally, the expression for the PEM fuel cell current in the p-th iteration, obtained through the proposed iterative procedure, is given by the following form:

$$I_{\text{fc}(p)} = \frac{b_1 g}{a_4} \Theta_{(p)}$$

$$I_{\text{fc}(p)} = \frac{b_1 g}{a_4} W \left(\frac{a_4}{b_1 g} \left(\frac{\left(1 - m \cdot I_{\text{fc}(p-1)}\right)^{\beta\left(\frac{g - h I_{\text{fc}(p-1)}}{I_{\text{fc}}}\right)} \cdot I_{\text{fc}(p-1)}^{-bh I_{\text{fc}}}}{e^{a_3 + a_5 I_{\text{fc}(p-1)}^2 + a_6 I_{\text{fc}(p-1)}^{3.5}}} \right)^{\frac{1}{b_1 g}} \right). \tag{5.15}$$

However, in the literature, two additional variants of the iterative solution procedure have been developed as extensions of a previously established model. These methods introduce greater flexibility by allowing different configurations of variables and arguments within the Lambert W function, thereby enhancing the algorithm's adaptability to specific engineering conditions.

Namely, the initial Eq. (5.1) can be rearranged into the following form:

$$e^{a_3 + a_4 I_{\text{fc}} + a_5 I_{\text{fc}}^2 + a_6 I_{\text{fc}}^{3.5}} = I_{\text{fc}}^{b(g - h I_{\text{fc}})} (1 - m \cdot I_{\text{fc}})^{\beta(g - h I_{\text{fc}})}; \tag{5.16}$$

which leads to the equation presented below:

$$\frac{7 a_6 I_{\text{fc}}^{3.5}}{2 b_1 g} e^{\frac{7 a_6 I_{\text{fc}}^{3.5}}{2 b_1 g}} = \frac{7 a_6}{2 b_1 g} \left(e^{-(a_3 + a_4 I_{\text{fc}} + a_5 I_{\text{fc}}^2)} I_{\text{fc}}^{-bh I_{\text{fc}}} (1 - m \cdot I_{\text{fc}})^{\beta(g - h I_{\text{fc}})} \right)^{\frac{7}{2 b_1 g}}; \tag{5.17}$$

that is, in the form of the Lambert W function:

$$\theta \cdot e^{\theta} = H_{\text{First - App}} \tag{5.18}$$

where

$$\theta = \frac{7 a_6 I_{\text{fc}}^{3.5}}{2 b_1 g},$$

$$H_{\text{First - App}} = \frac{7a_6}{2b_1g}\left(e^{-\left(a_3+a_4I_{fc}+a_5I_{fc}^2\right)}I_{fc}^{-bhI_{fc}}(1 - m \cdot I_{fc})^{\beta(g-hI_{fc})}\right)^{\frac{7}{2b_1g}}. \tag{5.19}$$

By applying the Lambert W function, θ can be expressed as:

$$\theta = W\left(H_{\text{First - App}}\right). \tag{5.20}$$

Following the previously derived expression, the new iterative value of the current is obtained as:

$${}^{(1)}I_{fc} = \left(\frac{2b_1g \cdot W\left(H_{\text{First - App}}\right)}{7a_6}\right)^{\frac{2}{7}}. \tag{5.21}$$

Furthermore, starting from the initial model Eq. (5.1), expressed in the form of an exponential–polynomial relationship, an algebraic transformation can be performed to obtain the following expression:

$$\frac{2a_5I_{fc}^2}{b_1g}e^{\frac{2a_5I_{fc}^2}{b_1g}} = \frac{2a_5}{b_1g}\left(e^{-\left(a_3+a_4I_{fc}+a_6I_{fc}^{3.5}\right)}I_{fc}^{-bhI_{fc}}(1 - m \cdot I_{fc})^{\beta(g-hI_{fc})}\right)^{\frac{2}{b_1g}}. \tag{5.22}$$

Following the general form of the Lambert W function, the previous relation can be expressed as:

$$\theta \cdot e^{\theta} = H_{\text{Second - App}}, \tag{5.25}$$

where

$$\theta = \frac{2a_5I_{fc}^2}{b_1g},$$

$$H_{\text{Second - App}} = \frac{2a_5}{b_1g}\left(e^{-\left(a_3+a_4I_{fc}+a_6I_{fc}^{3.5}\right)} \cdot I_{fc}^{-bhI_{fc}} \cdot (1 - m \cdot I_{fc})^{\beta(g-hI_{fc})}\right)^{\frac{2}{b_1g}}. \tag{5.23}$$

By applying the Lambert W function, θ in this case can be expressed as:

$$\theta = W\left(H_{\text{Second - App}}\right). \tag{5.24}$$

Based on the derived expression, the new iterative value of the current is expressed in closed form:

$${}^{(1)}I_{fc} = \left(\frac{b_1g \cdot W\left(H_{\text{Second - App}}\right)}{2a_5}\right)^{\frac{1}{2}}. \tag{5.25}$$

Therefore, there are different variants of applying the iterative Lambert W equation to solve the current–voltage characteristics of PEMFCs. Each variant has its advantages, disadvantages, and applicability efficiency depending on the values of the PEMFC block parameters. However, it is important to emphasize that PEMFC cells are inherently nonlinear, and various iterations of the Lambert W procedure can be effectively applied to solve them.

However, all the described approaches are applicable for low current values, i.e., for operating regimes close to open-circuit conditions. For high current values, the oscillatory iterative procedure may reach values exceeding the maximum current of the PEMFC, which, in mathematical terms, leads to the occurrence of complex numbers.

5.3.2 Main Part of the Realized MATLAB Code for Current-Voltage PEMFC Calculation

In this section, excerpts of the MATLAB code are provided for all three approaches to applying the Lambert W iterative procedure for determining the current–voltage characteristics of PEMFCs.

First approach

```
for current_iteration=1:max_iteration

FF=exp((a3+a5*initial^2+a6*initial^3.5))/(((1-m*initial)^(betaa*(g-
h*initial)))*initial^(-b*h*initial));
    GG=1/FF;
    HH=a4/(b1*g)*GG^(1/(b1*g));
    y=lambertw(HH);
    current=b1*g*y/a4;

FF=exp((a3+a5*current     ^2+a6*current     ^3.5))/(((1-m*current)^(betaa*(g-
h*current)))*current ^(-b*h*strujaa));
    GG=1/FF;
    HH=a4/(b1*g)*GG^(1/(b1*g));
    error=abs(HH-y*exp(y));
    if abs(real(error))<criteria
       break
    end

end
```

Second approach

```
for current_iteration=1:max_iteration;
```

```
FF=exp((a3+a4*initial+a5*initial^2))/((((1-m*initial)^(betaa*(g-
h*initial)))*initial^(-b*h*initial));
    GG=1/FF;
    HH=7*a6/(2*b1*g)*GG^(7/(2*b1*g));
    y=lambertw(HH);
    current=(2*b1*g*y/(7*a6))^(2/7);

FF=exp((a3+a4*current+a5*current          ^2))/((((1-m*current)^(betaa*(g-h*
current)))*current ^(-b*h*current));
    GG=1/FF;
    HH=7*a6/(2*b1*g)*GG^(7/(2*b1*g));
    error=abs(HH-y*exp(y));
    if abs(real(error))<criteria
       break
    end

end
```

Third approach

```
for current_iteration=1:max_iteration
                FF=exp((a3+a4*initial+a6*initial^3.5))/((((1-m*initial)^(betaa*(g-
h*initial)))*initial^(-b*h*initial));
    GG=1/FF;
    HH=2*a5/(b1*g)*GG^(2/(b1*g));
    y=lambertw(HH);
    current=(b1*g*y/(2*a5))^(1/2);
            FF=exp((a3+a4*current+a6*current^3.5))/((((1-m*current)^(betaa*(g-
h*current)))*current^(-b*h*current));
    GG=1/FF;
    HH=2*a5/(b1*g)*GG^(2/(1*b1*g));
    error=abs(HH-y*exp(y));
    if abs(real(error))<criteria
       break
    end

end
```

5.4 Numerical Results

To demonstrate the applicability of the derived solutions, a BSC 500W PEMFC with
the following parameters was considered.

% FC stack type BCS 500W

T = 333.

$Ma = 64.$

$l = 178*10^{-4}$

$Jmax = 0.469.$

$Tfc = 333.$

$Ph = 1.$

$Po = 0.2095.$

$N = 32.$

$ksi1 = -1.176591336.$

$ksi2 = 3.496528*10^{-3}$

$ksi3 = 5.8319*10^{-5}$

$ksi4 = -0.000192897.$

$lamda = 21.324205865.$

$Rc = 0.000146406.$

$betaa = 0.016140539.$

As an example, the second iterative procedure was examined. Figure 5.1 shows the current–voltage characteristics obtained by applying this iterative procedure, together with the measured characteristics of this solar cell. Figure 5.2 shows the convergence characteristics for all selected voltage values. There is a clear and evident match between the measured and simulated curves, as well as the convergence behavior, especially for low current values.

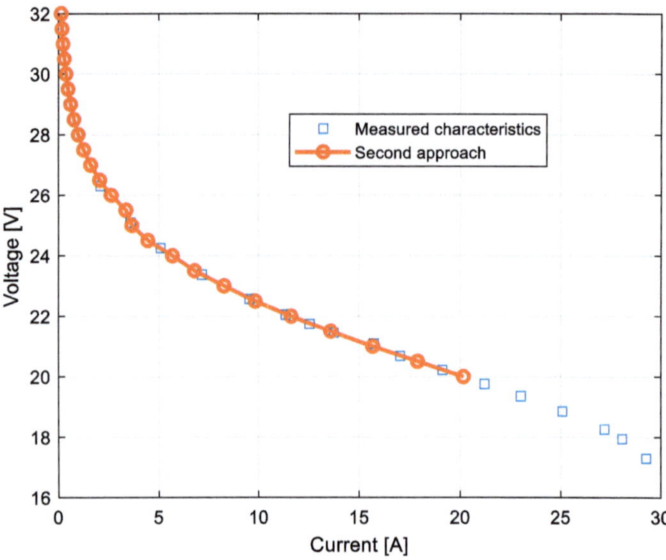

Fig. 5.1 Current–voltage characteristics of the BSC 500W PEMFC

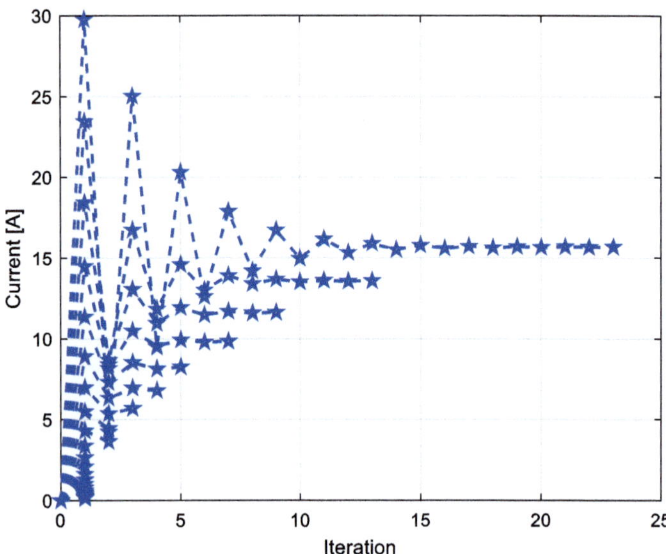

Fig. 5.2 Convergence characteristics

References

1. Pourrahmani H, Yavarinasab A, Siavashi M, Matian M, Van Herle J (2022) Progress in the proton exchange membrane fuel cells (PEMFCs) water/thermal management: from theory to the current challenges and real-time fault diagnosis methods. Energy Rev 1(1):100002. https://doi.org/10.1016/j.enrev.2022.100002
2. Manoharan P, Ravichandran S, Kavitha SS et al (2024) Parameter characterization of PEM fuel cell mathematical models using an orthogonal learning-based GOOSE algorithm. Sci Rep 14:20979. https://doi.org/10.1038/s41598-024-71223-7
3. Issa M, Abd Elaziz M, Selem SI (2025) Enhanced hunger games search algorithm that incorporates the marine predator optimization algorithm for optimal extraction of parameters in PEM fuel cells. Sci Rep 15:4474. https://doi.org/10.1038/s41598-025-87695-0
4. Vujošević S, Micev M, Ćalasan M (2025) Enhancing PEMFC parameter estimation: a comparative literature review and application of the Walrus optimization algorithm and its hybrid variants. J Renew Sustain Energy 17:024103. https://doi.org/10.1063/5.0245455
5. Ćalasan M, Abdel Aleem SHE, Hasanien HM, Alaas ZM, Ali ZM (2023) An innovative approach for mathematical modeling and parameter estimation of PEM fuel cells based on iterative Lambert W function. Energy 264:126165. https://doi.org/10.1016/j.energy.2022.126165
6. Ćalasan M, Micev M, Hasanien HM, Abdel-Aleem SHE (2024) PEM fuel cells: two novel approaches for mathematical modeling and parameter estimation. Energy 290:130130. https://doi.org/10.1016/j.energy.2023.130130
7. Ćalasan M (2025) Two iterative approaches using the Lambert W function for solving the Mann model in PEMFC: a short communication. Int J Hydrogen Energy 119:173–177. https://doi.org/10.1016/j.ijhydene.2025.03.245

Chapter 6
Nonlinear Diode-Resistance Circuits

This chapter is divided into several parts. The first part provides basic information about diode (D)—resistance (R) circuits. The second part discusses the mathematical modeling of D-R circuits, as well as realized MATLAB code. The third part presents the numerical results.

6.1 Basic Information About Diode (D)—Resistance (R) Circuits

The diode-resistor (D-R) circuit represents one of the most fundamental and at the same time highly significant configurations in the field of electrical and electronic circuits [1–3]. In this structure, a semiconductor diode is connected in series with a resistor, and the combination is supplied by either a voltage or current source. Despite its structural simplicity, the circuit exhibits pronounced nonlinear behavior, primarily due to the exponential dependence of the diode voltage on the current flowing through it.

In engineering practice, the diode-resistor (D-R) circuit is often regarded as a simple solution for nonlinear tasks involving current and voltage regulation. In many applications, this configuration enables the realization of functionalities that would otherwise require considerably more complex circuit implementations. The key feature of this circuit lies in its strongly nonlinear current-voltage relationship, arising from the exponential response of the diode and its interaction with the series resistance. Furthermore, in the D-R configuration, the voltage drop across the components and the resulting current are directly influenced by the diode's characteristics-such as saturation current, ideality factor, and thermal conditions-as well as by the value of the series resistor. This nonlinear dependence renders the analysis of the circuit highly complex, particularly in applications that demand high modeling accuracy, fast numerical processing, and robustness under varying operating conditions.

© The Author(s), under exclusive license to Springer Nature Switzerland AG 2026
M. Ćalasan and S. Vujošević, *Analytical Solutions of Nonlinear Power System Models Using the Lambert W Function*, SpringerBriefs in Electrical and Computer Engineering,
https://doi.org/10.1007/978-3-032-09579-4_6

The D-R circuit is frequently used as a fundamental functional block in systems where control of electrical parameters is required without the inclusion of active components. Its application spans various industries—from power electronics and automation, to lighting technology and protective circuitry.

D–R circuits have a wide range of practical applications in both electronic and power systems, particularly in scenarios that require control of nonlinear current–voltage relationships, energy regulation, output parameter stabilization, and protection of components under adverse operating conditions. Their nonlinear behavior does not represent a limitation, but rather a key functional advantage—provided the circuits are properly modeled and integrated into the system.

Among the most significant applications are the following:

- Photovoltaic systems: D-R circuits are used for modeling and analyzing the series resistance of solar cells, particularly in scenarios such as partial shading or elevated operating temperatures. The use of an accurate D-R model is essential for calculating the current-voltage (I-V) characteristics, optimizing MPPT algorithms, and maximizing energy conversion efficiency.
- Battery charging: In lithium-ion and other rechargeable battery technologies, D-R circuits enable controlled current limitation during the charging process, thereby improving efficiency and extending the service life of battery cells. When combined with control algorithms, they allow precise regulation of energy flow.
- LED lighting: D-R circuits provide current stabilization and limit overvoltage effects in LED power supplies. Precise current control enhances luminous efficiency and significantly reduces thermal stress on light-emitting components.
- Diode bridges and rectifiers: In both DC and AC power systems, the series resistance within D-R branches affects the rectified voltage profile, power dissipation, and transmission quality. Accurate modeling contributes to improved thermal and electrical stability.
- Voltage reference and stabilization sources: D-R circuits are applied in band-gap references and Zener-based regulators, where the series resistance enables fine adjustment of the operating point and improves temperature stability.
- Telecommunication systems and high-frequency oscillators: D-R structures are utilized in the design of communication circuits, where voltage stability is essential for minimizing noise and nonlinear distortion. They are also implemented in high-precision amplifiers and oscillators used for signal processing.
- Biomedical electronics: These circuits are applied in the development of low-voltage reference sources and programmable amplifiers, where high reliability and stability at very low voltages are critical.
- Protective electronic circuits and sensors: D-R circuits are integrated into input protection branches, often with antiparallel-connected diodes, to suppress overvoltage caused by transients or electrical pulses in both polarities. This application is particularly important in measurement and communication systems exposed to electromagnetic interference.

Due to its minimal component count and inherently robust behavior, the D-R circuit is often the preferred choice in designs where simplicity, cost-efficiency, and reliability are primary requirements. Although structurally simple, the behavior of this circuit necessitates precise mathematical modeling, especially in the design of systems operating over wide voltage ranges, under varying temperature conditions, or with high-frequency response requirements.

6.2 Diode-Resistor Circuits

This section consists of two parts. The first part presents the mathematical background of the D-R circuit, and the second part provides the implemented MATLAB code for solving this circuit.

6.2.1 Mathematical Modeling of D-R Circuits

The standard topology of the diode-resistor (D-R) circuit is presented in Fig. 6.1. The considered configuration includes a diode and a resistor connected in series, along with a voltage source that may be either direct current (DC) or alternating current (AC), depending on the specific system requirements. The diode conducts current only in one direction (under forward bias) and blocks it in the reverse direction. Its behavior in the forward-conduction mode is described by the Shockley diode equation, while the resistor establishes a linear relationship between current and voltage. The combination of these two elements results in a nonlinear differential relationship that requires dedicated mathematical treatment.

The voltage relationship that describes the behavior of the diode–resistor (D–R) circuit can be expressed in the following form:

$$V_S = V_D + RI. \tag{6.1}$$

Fig. 6.1 Standard configuration of the diode–resistor (D–R) circuit

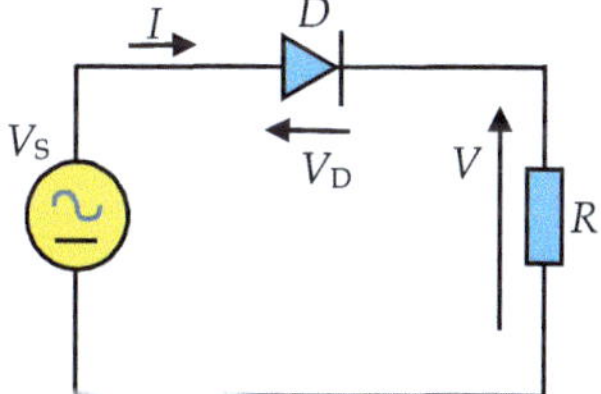

The current in the circuit is a function of the voltage across the diode and is expressed by the following equation:

$$I = I_0\left(e^{\frac{V_D}{aV_{\text{th}}}} - 1\right),\tag{6.2}$$

where

- V_D—voltage across the diode [V],
- I—current through the circuit [A],
- R—load resistance [Ω],
- V_S—supply voltage [V],
- I_0—diode saturation current [A],
- a—diode ideality factor,
- V_{th}—thermal voltage across the diode [V], calculated as: $V_{\text{th}} = kBT/q$
- q—elementary charge (1.602×10^{-19} C).
- kB—Boltzmann constant (1.38×10^{-23} J/K), and
- T—absolute temperature [K].

Equation (6.1) has an analytical solution expressed in terms of the Lambert W function, given in the following form:

$$I = \frac{aV_{\text{th}}}{R} W\left(\frac{RI_0}{aV_{\text{th}}} e^{\frac{V_S + RI_0}{aV_{\text{th}}}}\right) - I_0,\tag{6.3}$$

where W is the solution to the Lambert W equation.

However, there are scientific problems indicating that the value of the argument of the Lambert W function may exceed the maximum allowable limit, which is approximately 10^{323}. This represents a limitation in the applicability of the function in certain numerical analyses.

The previously mentioned limitation in the applicability of the Lambert W function, which arises at certain values of the supply voltage or circuit parameters, can be overcome by employing the so-called g-function.

This function represents a logarithmic transformation of the Lambert W function, enabling numerically stable computation even under extreme values of the argument, where the standard W function becomes inapplicable. As a result, the g-function provides a reliable solution for modeling the current in the D-R circuit over a wider range of voltage and temperature conditions.

The application of the *g*-function enables a precise analytical expression for the current in the D-R circuit, with the resulting formulation referred to as the *g*-exact solution [4]:

$$I = \frac{aV_{th}}{R}\left(\frac{V_S + RI_0}{aV_{th}} + \log\left(\frac{RI_0}{aV_{th}}\right) - g(x)\right) - I_0, \tag{6.4}$$

where

$$x = \frac{V_S + RI_0}{aV_{th}} + \log\left(\frac{RI_0}{aV_{th}}\right). \tag{6.5}$$

6.2.2 MATLAB Code for Current Calculation in D-R Circuits

The MATLAB code below enables the determination of the D-R circuit current, as well as diode voltage.

```matlab
% define circuit parameters

dt=1e-4

time=0:1e-4:0.04

for counter = 1:1:length(time)
    Vs = Vm*sin(2*pi*50*time(counter));
    alfa=(Vs+R*IS)/(a*Vth)+log(R*IS/(a*Vth));
    alfapoint(counter)=alfa;
        if alfa<-exp(1)
            y0=alfa;
        end
        if alfa>exp(1)
            y0=log(alfa);
        end
        if alfa<exp(1) && alfa>=-exp(1)
            y0=-exp(1)+(1+exp(1))/(2*exp(1))*(alfa+exp(1));
        end
        epsilon=10^-100;
        for iter=1:100
```

```
                y1=y0-(2*(y0+exp(y0)-alfa)*(1+exp(y0)))/(2*(1+exp(y0))^2-
(y0+exp(y0)-alfa)*exp(y0));
            if abs(y1-y0)<epsilon
               break
            end
            y0=y1;
         end
      gfunction=(y1);
            currentG(counter)=a*Vth/R*((Vs+R*IS)/(a*Vth)+log(R*IS/(a*Vth))-
gfunction)-IS;
      Vdiode(counter)=Vs-R*currentG(counter);
      end

figure(3)

plot(time,currentG,'b','Linewidth',2)

hold on

grid on

box on

xlim([0 0.04])

xlabel('Time [s]')

ylabel('I [A]')

figure(4)

plot(time,Vdiode,'b','Linewidth',2)

hold on

grid on

box on

xlim([0 0.04])

xlabel('Time [s]')

ylabel('V_D [V]')
```

figure(5)

plot(time,alfapoint,'b','Linewidth',2)

hold on

grid on

box on

xlim([0 0.04])

xlabel('Time [s]')

ylabel('\alpha')

6.3 Numerical Results

In order to validate the aforementioned theoretical considerations, this section presents numerical results obtained for different voltage values. The calculations were based on the characteristics of the power engineering MUR diode, which has the following parameters:

$I_S = 0.00307.$

$a = 6.67203.$

$T = 273.15 + 22;$

$q = 1.60219*10^{-19};$

$kB = 1.38062*10^{-23};$

$V_{th} = kB*T/q;$

Figure 6.2 shows the waveforms of the circuit current and the argument of the g-function for different supply voltage values, with a circuit resistance of 200 Ω. It can be observed that the value of the argument is extremely high, reaching several thousand. This confirms that, for these voltages, the modeling approach using the Lambert W function would not be applicable. Specifically, if the Lambert W approach were used, then for these real circuit parameters, the argument value would be an exponential function whose exponent is several thousand, which is numerically unacceptable according to the IEEE standard.

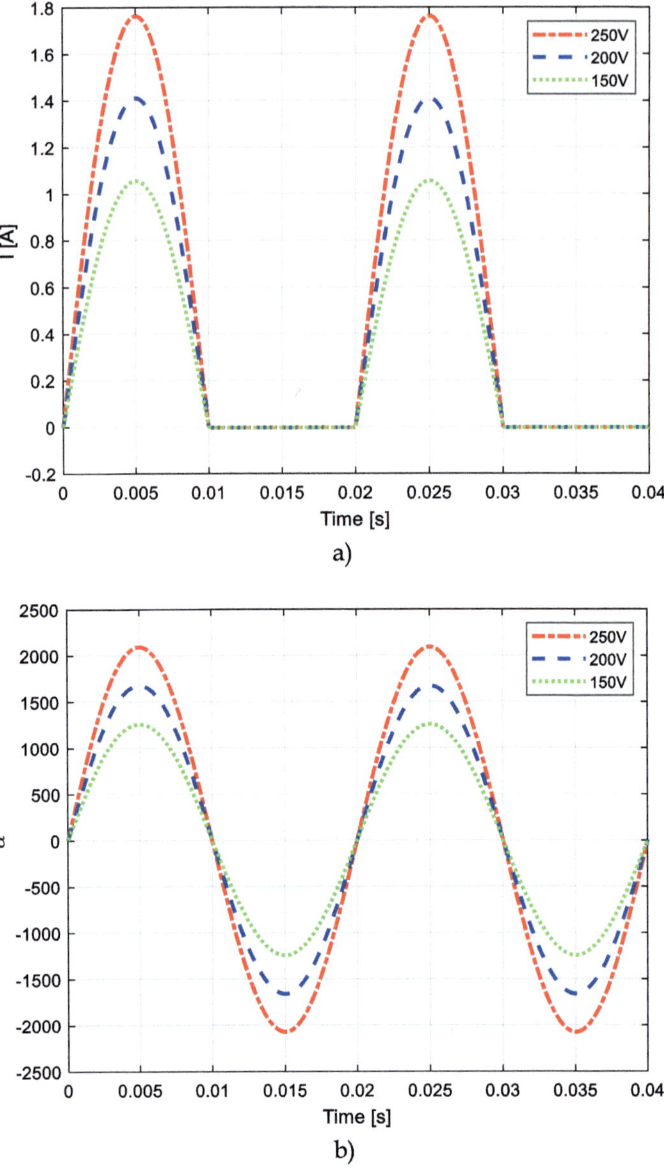

Fig. 6.2 D-R circuit **a** current, and **b** argument of *g*-function versus time

References

1. Banwell TC, Jayakumar A (2000) Exact analytical solution for current flow through diode with series resistance. Electron Lett 36(4):291–292. https://doi.org/10.1049/el:20000301
2. Ortiz-Conde A, García-Sánchez FJ (1992) Approximate analytical expression for equation of diode with series and shunt resistances. Electron Lett 28(21):1964–1965. https://doi.org/10.1049/el:19921259
3. Gázquez JA, Fernández-Ros M, Torrecillas B, Carmona J, Novas N (2021) New approximate analytical solution of the diode-resistance equation. AEU Int J Electron Commun 133:153665. https://doi.org/10.1016/j.aeue.2021.153665
4. Calasan M (2025) Diode-resistance circuit—novel exact and approximate solutions based on the g-function approach. AEU Int J Electron Commun 193:155744. https://doi.org/10.1016/j.aeue.2025.155744

Chapter 7
Single-Phase Diode Rectifiers

This chapter is divided into several parts. The first part provides basic information about single-phase rectifiers. The second part discusses the mathematical modeling of its, while the third part presents the numerical results.

7.1   Basic Information About Single-Phase Rectifiers

Diode rectifier circuits occupy a central place in the processes of converting alternating current (AC) to direct current (DC), particularly in devices that require a stable and reliable DC power source [1–3]. Due to their simple construction and reliable operation, such configurations are found in a wide range of electronic and power engineering systems—from mains power supplies and chargers to sensor and control circuits.

Diode rectifier circuits are used in numerous engineering and industrial applications, including:

- Power supply for electronic devices: The most common use of rectifier circuits is in converting AC mains voltage into a stable DC voltage for powering loads (computers, televisions, wireless routers, etc.).
- Battery chargers: In charging systems (e.g., lithium-ion batteries), rectifier circuits provide a controlled direct current flow necessary for safe charging.
- Power converters and inverters: Serving as the input or output stage in AC/DC conversions, including solar inverters and uninterruptible power supplies (UPS).
- Measurement and control: In low-voltage applications, diode rectifier circuits are used for signal detection, forming reference values, and passive voltage regulation.
- Telecommunication systems: Supplying stable power to sensitive components (amplifiers, filters, modems) through compact and reliable AC/DC rectifiers.

© The Author(s), under exclusive license to Springer Nature Switzerland AG 2026
M. Ćalasan and S. Vujošević, *Analytical Solutions of Nonlinear Power System Models Using the Lambert W Function*, SpringerBriefs in Electrical and Computer Engineering,
https://doi.org/10.1007/978-3-032-09579-4_7

- Protective and surge suppression circuits: As part of input branches that prevent reverse current flow and protect sensitive components from overvoltage and current surges.
- LED lighting and industrial electronics: Supplying DC voltage for LED sources and control circuits under various operating conditions.

In addition to the above, diode rectifier circuits are also used in the following technically specific structures:

- Phase-locked loops (PLL): Used to generate and stabilize reference voltages in phase-locking circuits, ensuring high accuracy in signal synchronization and processing.
- Implementation of linear resistance in active circuits: In modern analog structures, they are used as substitutes for ideal current sources, enabling improved linearity and design flexibility.
- Bandgap and temperature-independent voltage references: Due to their stability, rectifier circuits are a key part of reference voltage sources in sensitive electronic applications.

In their basic form, the diode functions as a current director, allowing current to pass only during the positive half-cycle, while the series resistor determines the load and affects the output characteristics. Although these topologies are simple, they exhibit pronounced nonlinear behavior due to the exponential relationship between the current and voltage of the diode, which requires precise mathematical modeling—especially in applications that demand high accuracy and dynamic conversion process analysis.

The primary function of a rectifier diode is the conversion of AC voltage into DC voltage. In its simplest form, a rectifier circuit consisting of an AC source, a diode, and a load eliminates the negative half-cycles of the input signal—the portions of voltage below zero are blocked, resulting in half-wave rectification. While this configuration provides basic functionality, it has limited efficiency because it utilizes only one half-cycle.

To overcome the limitations of half-wave rectification, full-wave rectifiers are used, which convert both positive and negative half-cycles of the input AC signal into a unidirectional output. In practice, the full-wave rectifier function can be implemented in two basic topologies:

- **Bridge configuration (Graetz connection)**: Uses four diodes arranged in a bridge form, eliminating the need for a center-tap transformer. This configuration enables efficient rectification without special symmetry requirements for the transformer, making it suitable for a wide range of applications.
- **Center-tap transformer topology**: Uses two diodes connected to the secondary winding of a transformer with a center tap. During the positive and negative half-cycles, each diode conducts alternately, enabling full-wave rectification with only two rectifying elements.

Regardless of the topology, diode operation introduces certain undesirable effects into the system, including the generation of higher harmonics and deterioration of power quality, which may require additional filtering and compensation measures during the design process.

7.2 Mathematical Models of a Diode Rectifier Circuit

In this section, single-phase bridge rectifiers are described, as well as rectifiers with a center-tapped transformer.

7.3 Single-Phase Diode Bridge Rectifier

Figure 7.1a shows the bridge configuration of a single-phase diode rectifier with purely resistive load. This topology consists of four diodes connected in a bridge arrangement, with an alternating voltage source applied at the input, and a resistor implemented as the load at the output. During the positive half-cycle of the input voltage, diodes D_1 and D_4 are conducting, allowing the circuit to be completed through the load (Fig. 7.1b). In the negative half-cycle (Fig. 7.1c), the conducting role is taken over by diodes D_2 and D_3, through which current also flows.

It is important to note that the current through the load retains the same direction in both half-cycles, since in each case a conductive path is established that ensures unidirectional current flow to the consumer. In this way, the input alternating voltage is effectively rectified, resulting in a direct voltage across the load.

The expression describing the load current of this circuit can be written as follows:

$$I_R = I_0\left(e^{\frac{V_S - RI_R}{2V_{th}}} - 1\right) + I_0\left(e^{-\frac{V_S + RI_R}{2V_{th}}} - 1\right), \tag{7.1}$$

where

- V_S—source voltage [V],
- I_R—load current [A],
- R—load resistance [Ω],
- I_0—diode saturation current [A], and
- V_{th}—thermal voltage of the diode [V].

The solution of this equation can be written in the following form, using the Lambert W function:

$$I_R = \frac{2V_{th}}{R}W(x) - 2I_0$$

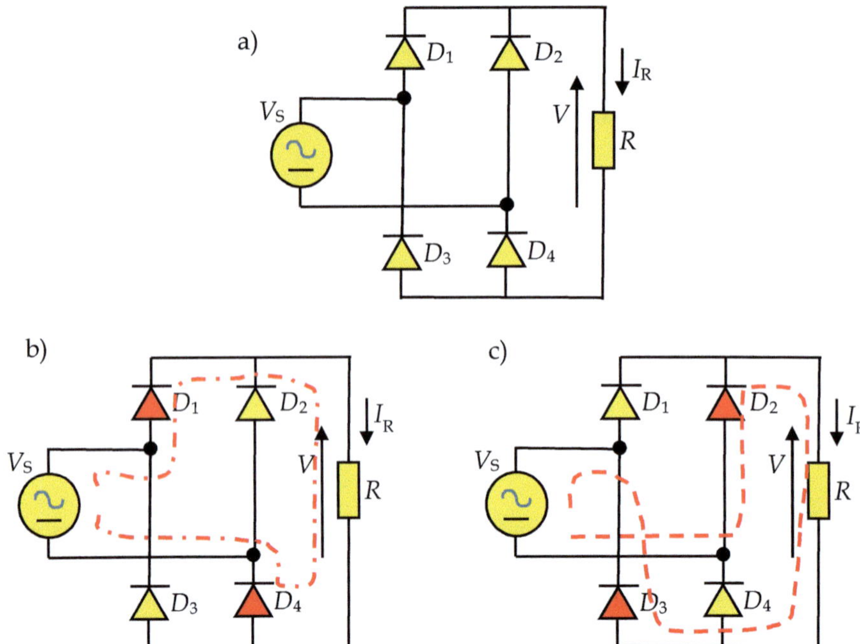

Fig. 7.1 **a** Single-phase diode bridge rectifier, **b** conduction of diodes D_1 and D_2 and **c** conduction of diodes D_3 and D_4

$$x = \frac{RI_0}{V_{th}} \cosh\left(\frac{V_S}{2V_{th}}\right) e^{\frac{RI_0}{V_{th}}}. \tag{7.2}$$

The load voltage is obtained by the simple application of Ohm's law, i.e., by multiplying the current flowing through the load by its resistance.

The source current is defined as the difference between the currents flowing through diode D_1 and diode D_2, which enables an accurate description of the current flow within the rectifier circuit:

$$I_S = I_{D1} - I_{D2}, \tag{7.3}$$

while

$$I_R = I_{D1} + I_{D2}$$
$$I_{D1} = I_0\left(e^{\frac{V_S - RI_R}{2V_{th}}} - 1\right)$$
$$I_{D2} = I_0\left(e^{-\frac{V_S + RI_R}{2V_{th}}} - 1\right). \tag{7.4}$$

Finally, the analytical expressions for the voltages across diodes D_1 and D_2 are obtained in the following form:

$$V_{D1} = \frac{V_S - RI_R}{2} = \frac{V_S}{2} - V_{th}W(x) + RI_0$$

$$V_{D2} = \frac{V_S + RI_R}{2} = \frac{V_S}{2} + V_{th}W(x) - RI_0. \tag{7.5}$$

It should be noted that the voltage waveforms across diode D_4 are identical to those recorded across diode D_1, while the voltage characteristics of diode D_3 are equivalent to those of diode D_2.

7.4 Single-Phase Rectifier with Center-Tapped Transformer

Figure 7.2a shows the configuration of a single-phase rectifier based on a transformer with a center-tapped secondary winding. In this design, the rectification is performed by only two diodes, which is the main difference compared to the bridge topology. During the positive half-cycle of the secondary voltage (Fig. 7.2b), the current flows through diode D_1, while during the negative half-cycle (Fig. 7.2c), diode D_2 conducts, with the transformer's center tap ensuring that the direction of current through the load remains unchanged. In this way, each diode conducts exclusively in one half-cycle, while a direct voltage and current are obtained at the output regardless of the polarity change of the input voltage.

Starting from the previously derived relation, expressions describing the currents and voltages of all elements in this rectifier configuration can be systematically obtained. In the further development of the analytical model, it is assumed that the transformer's primary voltage V_P has the same RMS value as the voltages across both secondary windings, i.e., $V_P = V_{S1} = V_{S2} = V_S$.

In that case, by denoting the transformer's secondary voltage as V_S, the expression for the load current can be written in the following form [4]:

$$I_R = I_0\left(e^{\frac{V_S - RI_R}{V_{th}}} - 1\right) + I_0\left(e^{-\frac{V_S + RI_R}{V_{th}}} - 1\right), \tag{7.6}$$

The solution of this equation can be written in the following form [5],

$$I_R = \frac{V_{th}}{R}W(x) - 2I_0$$

$$x = \frac{2RI_0}{V_{th}}\cosh\left(\frac{V_S}{V_{th}}\right)e^{\frac{2RI_0}{V_{th}}}. \tag{7.7}$$

Finally, the analytical expressions for the voltages across diodes D_1 and D_2 are obtained in the following form:

$$V_{D1} = V_S - RI_R = V_S - V_{th}W(x) + 2RI_0$$

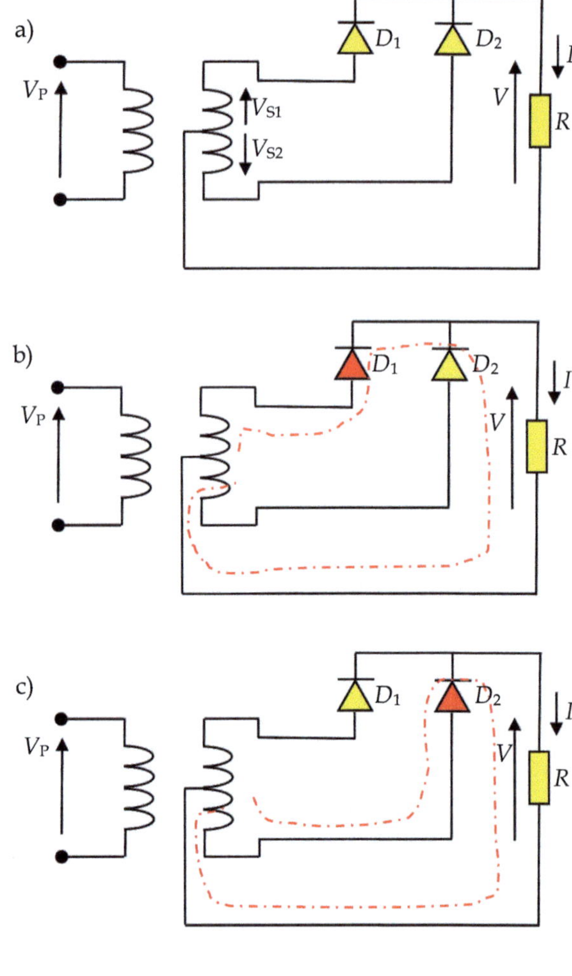

Fig. 7.2 **a** Single-phase rectifier with center-tapped transformer, **b** conduction of diode D_1 and **c** conduction of diode D_2

$$V_{D2} = V_S + RI_R = V_S + V_{th}W(x) - 2RI_0. \tag{7.8}$$

Therefore, the expression for the current and diode voltages of this type of single-phase rectifier is formulated using the Lambert W function.

7.5 Application of the g-function in Modeling Single-Phase Diode Rectifiers

As shown in Eq. (7.2) for the single-phase bridge rectifier and Eq. (7.7) for the single-phase rectifier with a center-tap transformer, the argument of the Lambert W function depends on the **cosh** function. Since **cosh** is expressed in terms of exponential functions, it becomes evident that, for certain values of its argument, the application of the Lambert W approach is not feasible due to numerical limitations. In such cases, the use of the g-function is more appropriate. Accordingly, the expressions for the currents and voltages of the single-phase bridge rectifier can be formulated in terms of the g-function as follows [4]:

$$I_R = \frac{2V_{th}}{R}(\alpha - g(\alpha)) - 2I_0, \tag{7.9}$$

where

$$\alpha = \log\left(\frac{RI_0}{V_{th}}\right) + \log\left(\cosh\left(\frac{V_S}{2V_{th}}\right)\right) + \frac{RI_0}{V_{th}}, \tag{7.10}$$

$$I_S = I_{D1} - I_{D2} = I_0\left(e^{\frac{V_S - RI_R}{2V_{th}}} - 1\right) - I_0\left(e^{-\frac{V_S + RI_R}{2V_{th}}} - 1\right), \tag{7.11}$$

$$
\begin{aligned}
V_{D1} &= \frac{V_S - RI_R}{2} \\
&= \frac{V_S}{2} - V_{th}\left(\log\left(\frac{RI_0}{V_{th}}\cosh\left(\frac{V_S}{2V_{th}}\right)\right) - g(\alpha)\right),
\end{aligned}
\tag{7.12}
$$

$$
\begin{aligned}
V_{D2} &= \frac{V_S + RI_R}{2} \\
&= \frac{V_S}{2} + V_{th}\left(\log\left(\frac{RI_0}{V_{th}}\cosh\left(\frac{V_S}{2V_{th}}\right)\right) - g(\alpha)\right).
\end{aligned}
\tag{7.13}
$$

Similarly, the expressions for the currents and voltages in a single-phase rectifier with a center-tapped transformer reduce to the following form:

$$I_R = \frac{V_{th}}{R}(\alpha - g(\alpha)) - 2I_0, \tag{7.14}$$

where

$$\alpha = \log\left(\frac{2RI_0}{V_{th}}\cosh\left(\frac{V_S}{2V_{th}}\right)\right) + \frac{2RI_0}{V_{th}}, \tag{7.15}$$

$$V_{D1} = V_S - RI_R$$

$$= V_S - V_{th}\left(\log\left(\frac{2RI_0}{V_{th}}\cosh\left(\frac{V_S}{V_{th}}\right)\right) - g(\alpha)\right), \qquad (7.16)$$

$$V_{D2} = V_S + RI_R$$

$$= V_S + V_{th}\left(\log\left(\frac{2RI_0}{V_{th}}\cosh\left(\frac{V_S}{V_{th}}\right)\right) - g(\alpha)\right). \qquad (7.17)$$

The preceding equations represent the application of the g-function in calculating the current of single-phase diode rectifiers.

7.6 MATLAB Code for Single-Phase Diode Rectifiers

The following section provides the MATLAB code for determining the load current and the diode voltages in a single-phase bridge rectifier.

```
point=1;

for time=0:0.0001:0.06

Ve=Vm*sin(w*time);

alfa=log(q*R*IS/(a*kB*T))+log(cosh(q*Ve/(2*a*kB*T)))+(q*R*IS/(a*kB*T));

if Ve>=0

alfa=log(q*R*IS/(a*kB*T))+q*Ve/(2*a*kB*T)+log((1+exp(-2*q*Ve/
(2*a*kB*T)))/2)+(q*R*IS/(a*kB*T));

end

if Ve<0

alfa=log(q*R*IS/(a*kB*T))-q*Ve/(2*a*kB*T)+log((1+exp(2*q*Ve/(2*
a*kB*T)))/2)+(q*R*IS/(a*kB*T));

end

alfapoint(point)=alfa;
        if alfa<-exp(1)
          y0=alfa;
        end
        if alfa>exp(1)
          y0=log(alfa);
        end
        if alfa<exp(1) && alfa>=-exp(1)
          y0=-exp(1)+(1+exp(1))/(2*exp(1))*(alfa+exp(1));
        end
        epsilon=10^-100;
```

```
        for iter=1:100
                y1=y0-(2*(y0+exp(y0)-alfa)*(1+exp(y0)))/(2*(1+exp(y0))^2-
(y0+exp(y0)-alfa)*exp(y0));
            if abs(y1-y0)<epsilon
               break
            end
            y0=y1;
        end

gfunction=(y1);

IR=2*a*kB*T /(q*R)*(alfa-gfunction)-2*IS;

IRpoint(point)=IR;

VAC=Ve/2+IR*R/2;

VAB=Ve/2-IR*R/2;

VACpoint(point)=VAC;

VABpoint(point)=VAB;

timepoint(point)=time;

Vinpoint(point)=Ve;

point=point+1;

end
```

7.7 Numerical Results

In order to validate the aforementioned theoretical considerations, this section presents numerical results for single phase diode bridge rectifier. The calculations were based on the characteristics of the MUR diode.

Figure 7.3 shows numerical results for various supply voltage values, with a load resistance of $100\,\Omega$. As the supply voltage increases, both the load current and the diode voltages rise accordingly, as evidenced by the results. Additionally, higher voltages lead to larger values of the g-function argument. In the presented examples, the argument reaches values of several thousands, indicating that the Lambert W function is numerically impractical for these voltage levels, thus necessitating the use of the g-function for accurate modeling.

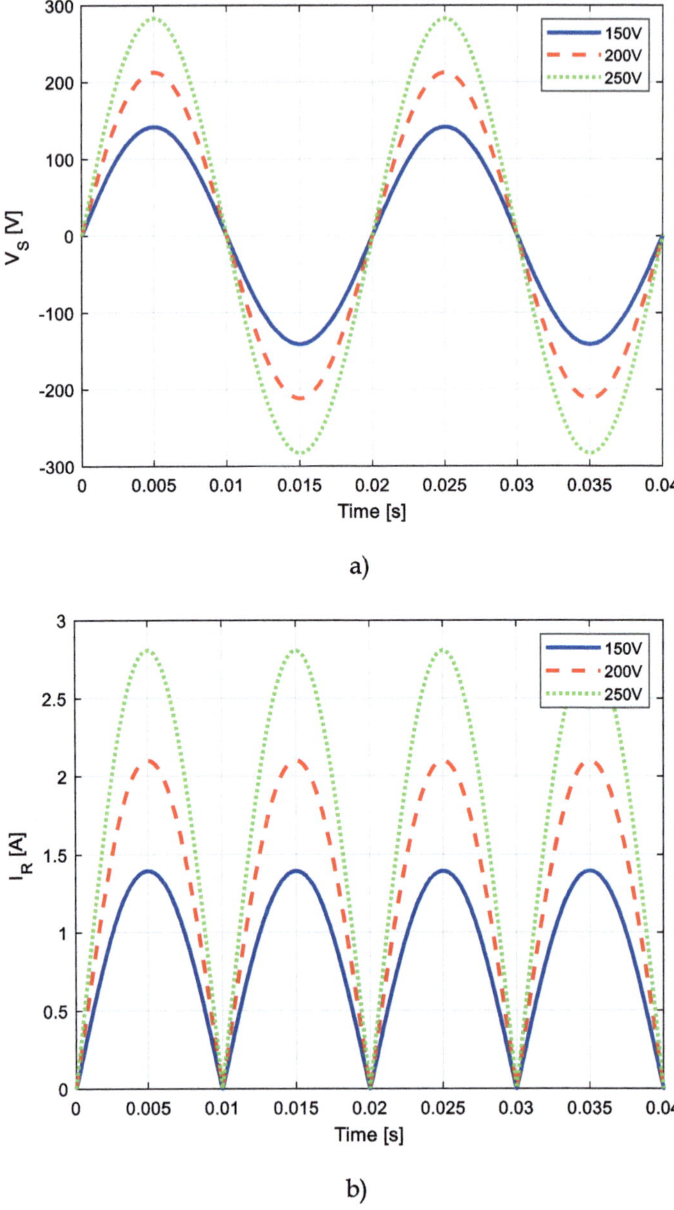

a)

b)

Fig. 7.3 Single-phase bridge rectifier: **a** supply voltage, **b** load current, **c** voltage across diode D_1, **d** voltage across diode D_2, and **e** argument of the g-function

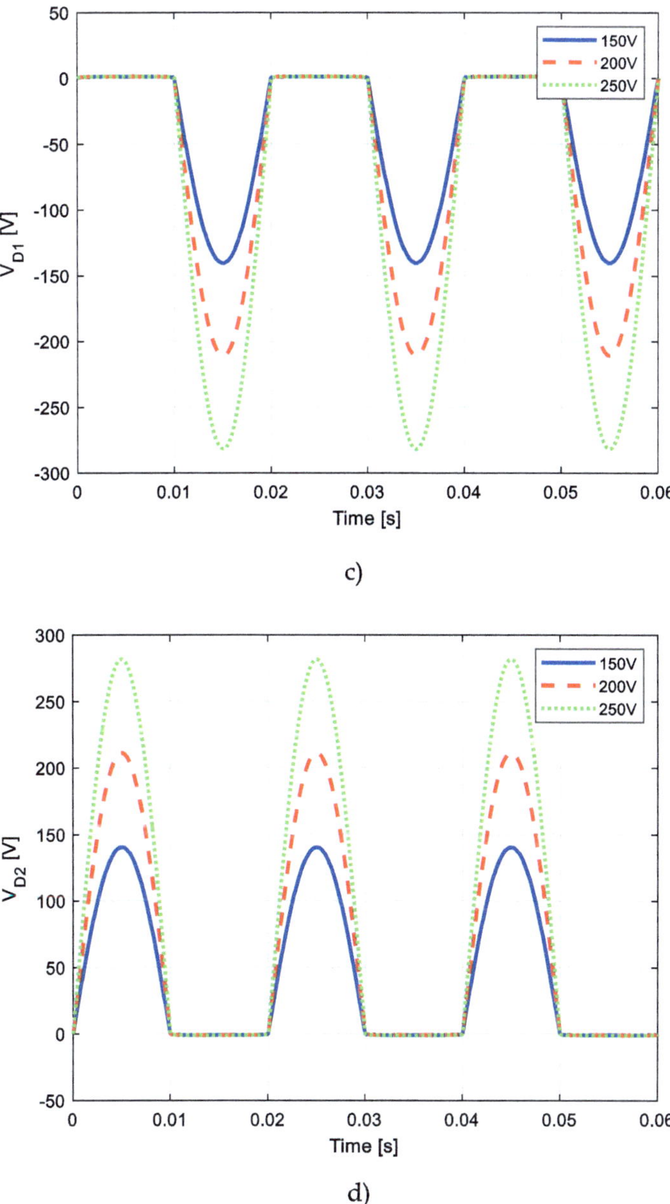

Fig. 7.3 (continued)

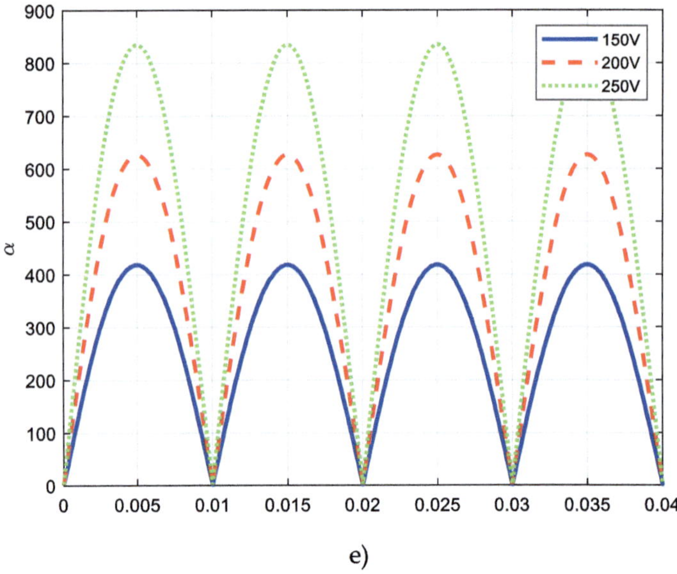

Fig. 7.3 (continued)

References

1. Behtash R et al (2022) Diode application handbook—fundamentals, characteristics, applications, 1st ed. Nexperia
2. Rashid M (2018) Power electronics handbook, 4th ed. Elsevier. https://doi.org/10.1016/C2016-0-00847-1
3. Asadi F (2023) Diode circuits. In: Analog electronic circuits laboratory manual, pp 1–33. https://doi.org/10.1007/978-3-031-25122-1_1
4. Calasan M (2025) Closed-form analytical solution of currents and voltages in single-phase diode rectifiers with a purely resistive load using the g-function approach. Measurement 256:118082. https://doi.org/10.1016/j.measurement.2025.118082
5. Assaid EM, Feddi EM, Aydi ME (2007) Exact analytical expressions of Gra¨etz bridge currents and voltages using Lambert W function. In: 14th IEEE international conference on electronics, circuits and systems, pp 278–282. https://doi.org/10.1109/ICECS.2007.4510984

Chapter 8
Analytical Modeling of Inductors' Air Gap Length Using Lambert W Function

This chapter consists of three parts. The first part provides basic information about power inductors. The second part describes the application of the Lambert W function in modeling the air gap length in inductors, while the third part presents a numerical example.

8.1 Basic Information About the Power Inductor

Power inductors represent passive electrical components with extensive application across contemporary power and automation systems [1]. They serve a fundamental function in power conversion circuits, where they contribute to the processes of energy transformation, voltage regulation, and electromagnetic noise filtering. Beyond their role in converters, power inductors are also integral to the operation of electric vehicles, power factor correction systems, rectifiers, DC-DC converters, and a wide array of other electronic assemblies [2].

In contrast to signal inductors, which are designed for processing low-power signals, power inductors are engineered to operate under medium and high power conditions. This entails substantial voltage and current levels typically encountered in power supplies, electric motors, drives, and industrial systems. As a result, these inductors are physically larger, heavier, and often constitute a limiting factor in terms of the size, weight, efficiency, and cost of an electronic device.

Consequently, their proper selection and sizing play a crucial role during the design phase.

The fundamental function of a power inductor is to store energy in the magnetic field generated by the flow of electric current through its winding. Owing to this property, power inductors enable:

M. Ćalasan and S. Vujošević, *Analytical Solutions of Nonlinear Power System Models Using the Lambert W Function*, SpringerBriefs in Electrical and Computer Engineering, https://doi.org/10.1007/978-3-032-09579-4_8

- controlled energy transfer,
- filtering of unwanted frequency components and electromagnetic interference (EMI),
- mitigation of transients during load changes or switching events,
- the formation of a stable and continuous output voltage and current.

On the other side, power inductors are key components in various automation subsystems, such as:

- power supplies for programmable logic controllers (PLCs),
- switching DC/DC and AC/DC converters,
- variable frequency drives (VFDs) for controlling asynchronous and synchronous motors,
- reactors and electromagnetic interference (EMI) filters,
- noise reduction and isolation systems.

In industrial applications where requirements for reliability, response, and efficiency are high, power inductors enable:

- stable and secure power supply to control and actuator units,
- compensation of overvoltage spikes and pulses,
- improvement of power quality for sensors, communication, and control modules.

Combined with other passive components (capacitors, resistors), power inductors are employed to form various types of filters (low-pass, dual-purpose, and multi-stage), thereby safeguarding sensitive electronics and enhancing system electromagnetic compatibility (EMC). Moreover, owing to their function as an "energy reservoir," they ensure a continuous flow of energy even during rapid state changes in the circuit.

Also, it must be emphasized that the overall efficiency of power circuits depends to a large extent on the efficiency of the inductors integrated into those circuits. Accordingly, controlling losses in magnetic components is essential for enhancing overall system efficiency. Furthermore, a properly designed and optimized power inductor makes a significant contribution to the stability, efficiency, and longevity of the entire system.

The construction of a power inductor typically consists of an electrical conductor (e.g., wire) wound into a coil around a core (Fig. 8.1). Inductors with high-permeability cores not only incur increased losses but also exhibit a lower level of magnetic-flux saturation, thereby reducing total inductance, which may be undesirable in certain applications. Therefore, in power-electronics applications, inductors with an air gap in the core are frequently used (see Fig. 8.1).

The introduction of an air gap significantly modifies the characteristics of the magnetic circuit:

- Increases the saturation current,
- Linearizes the B–H curve (relationship between magnetic flux density and magnetic field strength),
- Reduces the inductance.

Fig. 8.1 Power inductor

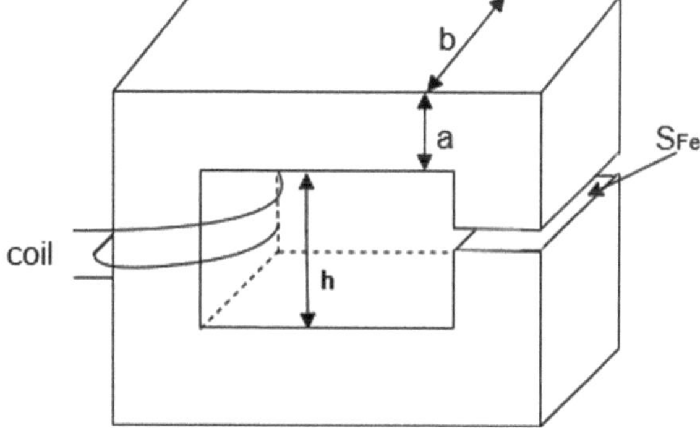

Designing an inductor with an air gap is highly complex because the behavior of the magnetic material is distinctly nonlinear. Figure 8.2 illustrates the difference between the B–H curve of a core without an air gap and one with an air gap.

However, a large air gap generates the so-called fringing flux (magnetic flux leakage). This effect has two consequences:

- The inductance increases due to the enlarged effective air-gap area, which reduces its reluctance.
- The fringing flux induces eddy currents on the conductor's surface near the gap, increasing losses and heating the windings.

The intensity of the fringing effect depends on:

- the dimensions of the air gap,

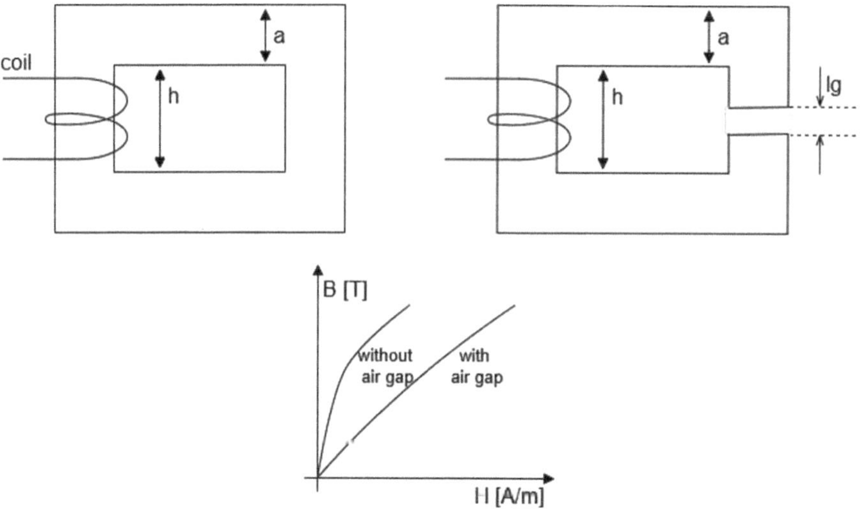

Fig. 8.2 Comparison of the B-H curve for the magnetic core without and with air-gap

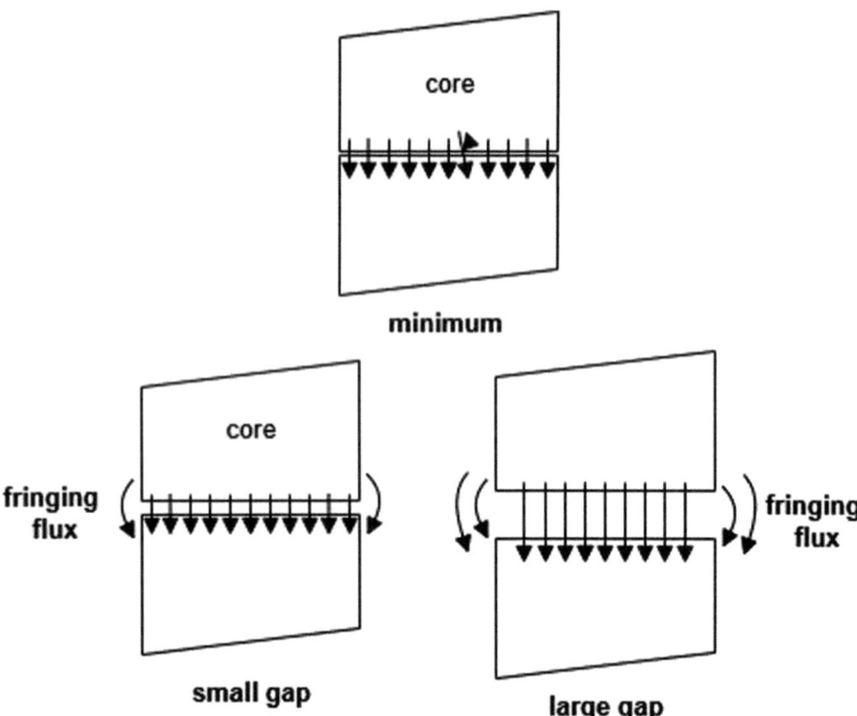

Fig. 8.3 Fringing flux at the gap

- the shape of the core poles,
- the size, geometry, and position of the windings.

Figure 8.3 depicts a graphical representation of the influence of the fringing effect.

The inductance of an iron-core inductor with an air gap, neglecting fringing effects, is given by the formula [3, 4]:

$$L = \mu_0 N^2 \frac{S_{Fe}}{l_g}, \tag{8.1}$$

where μ_0 is the magnetic permeability of vacuum, N the number of turns, l_g the length of the air gap, and S_{Fe} the core's cross-sectional area.

To account for the fringing effect, a correction factor FF is introduced:

$$FF = 1 + \frac{l_g}{a\pi} \ln\left(\frac{2h}{l_g}\right), \tag{8.2}$$

where h is leg height, and a is leg length.

In that case, the expression for inductance that includes the fringing effect can be written as:

$$L = \frac{\mu_0 N^2 S_{Fe}}{l_g} FF.$$

(8.3)

By substituting the expression for the correction factor FF from Eq. (8.2) into the inductance relation of Eq. (8.3), the complete expression for the inductance including the fringing effect is obtained:

$$L = \mu_0 N^2 \frac{S_{Fe}}{l_g}\left(1 + \frac{l_g}{a\pi}\ln\left(\frac{2h}{l_g}\right)\right).$$

(8.4)

This expression enables more accurate calculations when the influence of the fringing effect is pronounced, particularly for larger air gaps and at higher operating frequencies, where the "leakage" of magnetic flux beyond the core becomes increasingly significant. However, incorporating the fringing effect into the inductance formula inevitably introduces nonlinearity into the relationship between physical parameters and inductance, especially concerning the air-gap length.

In the expression that incorporates the fringing effect, it is evident that inductance no longer depends linearly on the air-gap length l_g. Specifically, the additional term, namely the logarithmic function, introduces a high degree of nonlinearity into the relationship between L and l_g, which significantly complicates inductor design when a precisely defined inductance value is required. This nonlinearity stems from the physical behavior of the magnetic field: as the air gap increases, magnetic flux increasingly "leaks" outside the core, thereby enlarging the effective cross-sectional area through which the flux passes. Consequently, the total magnetic reluctance of the core decreases more slowly than a simple linear relation would predict, resulting in a more gradual reduction of inductance.

Therefore, in practical applications, classical linear approximations prove inadequate, particularly for larger air-gap lengths or when stringent accuracy requirements are imposed (e.g., in high-efficiency power converters). In such circumstances, employing the inductance expression that incorporates the fringing effect is essential for obtaining an accurate inductor model.

8.2 Application of the Lambert W Function in Power Inductors Modeling with MATLAB Codes

The following section presents the methodology for applying the Lambert W function to solve the expression for l_g, along with a MATLAB implementation that enables engineers to easily apply this approach in real-world design projects.

8.2.1 Analytical Solution for Air-Gap Length

Equation (8.4), which defines the relationship between inductance and the geometric parameters of the core while accounting for the fringing effect, has a complex nonlinear form. In this expression, the air-gap length l_g appears both in the denominator and inside a logarithmic function, making the task of solving for l_g particularly challenging.

To obtain a closed-form analytical expression for the air-gap length, the equation can be transformed into a form suitable for solving using the Lambert W function. This function enables the inversion of nonlinear transcendental expressions in which the unknown variable appears both as a factor and as an exponent, precisely the case in the given relation.

In this context, Eq. (3.4), through appropriate algebraic manipulation, can be rewritten in the following form:

$$L = \mu_0 N^2 \frac{S_{Fe}}{l_g} + \frac{S_{Fe}}{a\pi} \ln\left(\frac{2h}{l_g}\right). \tag{8.5}$$

After substituting the expression for the effective core cross-sectional area $S_{Fe} = a{\cdot}b$, the relation takes on a simplified form:

$$\frac{L\pi}{\mu_0 N^2 b} = \frac{a\pi}{l_g} + \ln\left(\frac{2h}{l_g}\right). \tag{8.6}$$

After the corresponding mathematical steps, the following relation is obtained:

$$\frac{a\pi}{l_g} = \frac{a\pi}{2h} e^{\frac{L\pi}{\mu_0 N^2 b}} e^{-\frac{a\pi}{l_g}}. \tag{8.7}$$

It is evident that, after algebraic simplification, the relation takes the form:

$$W = xe^{-W}, \tag{8.8}$$

which represents the standard form suitable for solution using the Lambert W function. In this expression, $W = a\pi/l_g$, meaning it depends on the unknown air gap length, while x is a composite function of known quantities (both geometric and electrical):

$$x = \frac{a\pi}{2h} e^{\frac{L\pi}{\mu_0 N^2 b}}. \tag{8.9}$$

Based on the preceding derivation, and following Eqs. (8.8) and (8.9), the air gap length can finally be expressed in a closed analytical form as [4]:

$$l_g = \frac{a\pi}{W\left(\frac{a\pi}{2h} e^{\frac{L\pi}{\mu_0 N^2 b}}\right)}. \tag{8.10}$$

This form enables a closed and direct solution for the air gap length without the need for iterative methods.

If an analytical expansion of the solution is required, Lagrange's inversion theorem can be applied, yielding the following form for the air gap length:

$$l_g = \frac{a\pi}{\sum_{n=1}^{M} \frac{(-n)^{n-1}}{n!} \left(\frac{a\pi}{2h} e^{\frac{L\pi}{\mu_0 N^2 b}}\right)^n}. \tag{8.11}$$

However, when the parameter x exceeds the value of 3, which is a common case in real-world applications, it is advisable to use the expression given by Eq. (3.12), as the direct use of the closed form of the Lambert W function in this range may lead to reduced accuracy. In such circumstances, the air gap length can be determined using the following expression:

$$l_g = \frac{a\pi}{\ln\left(\frac{a\pi}{2h} e^{\frac{L\pi}{\mu_0 N^2 b}}\right) - \ln\left(\ln\left(\frac{a\pi}{2h} e^{\frac{L\pi}{\mu_0 N^2 b}}\right)\right) + \sum_{l=0}^{M} \sum_{m=1}^{M} \frac{(-1)^l \begin{bmatrix} l+m \\ l+1 \end{bmatrix}}{m!} \ln\left(\frac{a\pi}{2h} e^{\frac{L\pi}{\mu_0 N^2 b}}\right)^{-l-m} \ln\left(\ln\left(\frac{a\pi}{2h} e^{\frac{L\pi}{\mu_0 N^2 b}}\right)\right)^m}. \tag{8.12}$$

The expression obtained for larger values of the parameter x, as given by Eq. (8.12), represents a logarithmic expansion derived from the asymptotic form of the Lambert W function. This approach enables an analytical solution to the transcendental equation in domains where the function's argument is large, that is, when $x > 3$. Under these conditions, the asymptotic expansion provides a stable and accurate approximation for the air gap length, ensuring the reliability of the model even in regimes characterized by strong nonlinearity.

This expanded form offers:

- greater accuracy compared to simple logarithmic approximations,
- numerical stability for high values of x,
- improved control over precision depending on the number of terms M used in the summation.

Additionally, if a solution to the transcendental equation is approached using a method based on the STFT, it is possible to derive an alternative analytical form for the air gap length, as follows:

$$l_g = 2he^{-\frac{L\pi}{\mu_0 N^2 b}} \frac{\sum_{n=0}^{M+1} \frac{\left(\frac{a\pi}{2h} e^{\frac{L\pi}{\mu_0 N^2 b}}\right)^n}{n!}(M+1-n)^n}{\sum_{n=0}^{M} \frac{\left(\frac{a\pi}{2h} e^{\frac{L\pi}{\mu_0 N^2 b}}\right)^n (M-n)^n}{n!}}.$$

(8.13)

Accordingly, three analytical expressions for modeling the air gap length of an inductor have been derived using different approaches. Each of these methods has its advantages and limitations in terms of accuracy, computational efficiency, and practical applicability. Depending on the specific requirements of the inductor design and simulation process, an appropriate model can be selected.

8.2.2   MATLAB Code for Inductor Air-Gap Length Calculation

The MATLAB code below enables the determination of air gap length as a function of varying inductance values.

```matlab
clc
clear all
close all
format long
a = 35e-3;
b = 40e-3;
h = 120e-3;
N = 50;
mu0 = 4*pi*1e-7;
L = linspace(0.2e-3, 0.8e-3, 50); % [H]
l_g = zeros(size(L));
x = zeros(size(L));
for k = 1:numel(L)
        x(k) = (a*pi)/(2*h) * exp( (pi * L(k)) / (mu0 * N^2 * b) );
        l_g(k) = (a*pi) / lambertw( x(k) );
end
figure(1)
plot(L*1e3, l_g*1e3,'Linewidth',2)
```

grid on

xlabel('Inductance L [mH]')

ylabel('Air-gap length l_g [mm]')

box on

figure(2)

plot(L*1e3, x*1e3,'Linewidth',2)

grid on

xlabel('Inductance L [mH]')

ylabel('Argument x')

box on

8.3 Numerical Results

In order to validate the theoretical considerations and assess the practical applicability of the derived analytical expressions, this section presents numerical results obtained for different inductance values. The calculations were based on characteristic design parameters of the magnetic circuit, representative of typical applications in power electronics and automated systems:

- $a = 35$ mm (core leg length),
- $b = 40$ mm (cross-sectional width),
- $h = 120$ mm (core leg height),
- $N = 50$ (number of turns).

The examined configuration enables a realistic assessment of model behavior and allows testing of the expressions in a domain where nonlinear effects are prominent. Specifically, based on the given input data, the parameter x, which appears in all analytical formulations, reaches values significantly greater than 3. This indicates that, if one of the analytical methods is to be applied, it is advisable to use either the STFT-based approach or the asymptotic solution.

The variation of air gap length with inductance is presented in Fig. 8.4a, while Fig. 8.4b shows the corresponding behavior of the function argument.

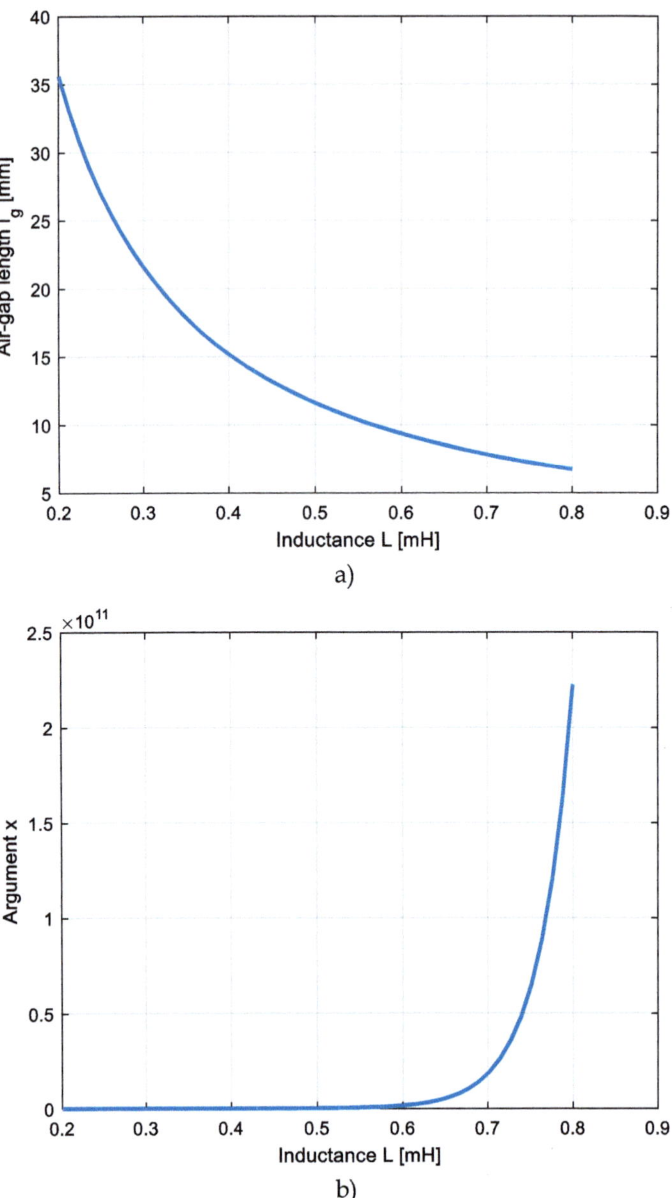

Fig. 8.4 Inductor **a** air gap length and **b** Lambert W argument change via inductance

References

1. Ravera A, Oliveri A, Lodi M, Beatrice C, Ferrara E, Fiorillo F, Storace M (2025) Modeling amorphous-core inductors up to magnetic saturation. IEEE Trans Power Electron 40(1):1563–1576. https://doi.org/10.1109/TPEL.2024.3463803
2. Zeng X, Chen W, Yang L, Chen Q, Huang Y (2024) An accurate calculation method for inductor air gap length in high power DC–DC converters. Sci Rep 14:3473. https://doi.org/10.1038/s41598-024-54194-7
3. Nedić AB, Simović MV, Lazarević ZM, Milić SD (2016) Implementation of minimisation techniques to construction optimisation of iron-core inductor. IET Electr Power Appl 10(1):9–17. https://doi.org/10.1049/iet-epa.2014.0446
4. Ćalasan M, Nedić A (2018) Experimental testing and analytical solution by means of Lambert W-function of inductor air gap length. Electr Power Compon Syst 46(7):852–862. https://doi.org/10.1080/15325008.2018.1488012

If you have any concerns about our products,
you can contact us on
ProductSafety@springernature.com

In case Publisher is established outside the EU,
the EU authorized representative is:
Springer Nature Customer Service Center GmbH
Europaplatz 3, 69115 Heidelberg, Germany

Printed by Libri Plureos GmbH
in Hamburg, Germany